Elastic Waves in Solids 2

Series Editors
Pierre-Noël Favennec† and Frédérique de Fornel

Elastic Waves in Solids 2

Radiation, Scattering, Generation

Tony Valier-Brasier
Daniel Royer

iSTE

WILEY

First published 2022 in Great Britain and the United States by ISTE Ltd and John Wiley & Sons, Inc.

ISTE Ltd
27-37 St George's Road
London SW19 4EU
UK

www.iste.co.uk

John Wiley & Sons, Inc.
111 River Street
Hoboken, NJ 07030
USA

www.wiley.com

Library of Congress Control Number: 2022931520

British Library Cataloguing-in-Publication Data
A CIP record for this book is available from the British Library
ISBN 978-1-78630-815-3

Contents

Preface

In Volume 1 (Royer and Valier-Brasier 2022), the propagation of elastic waves was studied in isotropic, anisotropic, piezoelectric or viscoelastic solids. The phenomena of reflection and transmission of bulk waves at one or more interfaces were analyzed in detail. A wide variety of elastic waves propagating on a free surface or at the interface between a solid and a fluid, or between two solids, were presented. Finally, the properties of various types of waves guided in most usual structures (plate, ribbon, cylinder) were established. In addition to these general developments, the objective of this second volume is to analyze the radiation, scattering and generation of elastic waves.

Plane waves are not physically realizable because, in practice, sources are of finite dimensions. Acoustic waves generated by a real device diverge, so that their amplitude decreases with the distance from the transmitter. The understanding of radiation from sources is therefore necessary to predict the acoustic field emitted by the transducers used in imaging and measurement systems. In Chapter 1, we start by studying the acoustic radiation in a fluid emitted by a vibrating surface having the shape of a disc or a spherical cup, respectively, modeling a planar or a focused transducer. Then, we study the radiation emitted by a multi-element antenna such as those used in ultrasound imaging systems. In the second section, we analyze the radiation of elastic waves, in an isotropic or anisotropic solid, by sources located on its surface. The generation of bulk and Rayleigh waves by a linear distribution of normal or tangential impulsive forces, modeling a seismic or a thermoelastic source, is examined. The last section is devoted to the radiation of an elementary spherical source embedded in an isotropic matrix.

Phenomena related to the scattering of elastic waves by various targets are developed in Chapter 2. The scattering theory accounts for different phenomena observed in Volume 1 (refraction, transmission, absorption) in a global way. The scattered waves propagate away from the target, possibly in preferred directions, even if the medium is isotropic. As in other areas of physics, scattering phenomena are characterized by a cross-section, which can be calculated in the case of cylindrical or spherical scatterers. In the first section, the target is a cylinder immersed in a perfect fluid; the incident wave is therefore longitudinal. The two examples studied explain some phenomena, such as the shadow behind a rigid cylinder in the geometric regime, or the existence of circumferential waves propagating around an elastic cylinder in the resonant regime. The second section is devoted to the scattering of elastic waves by a cylinder buried in a solid matrix, and the third section is devoted to the scattering by a sphere. In the case of a sphere, strong monopolar and dipolar resonances are highlighted. The multiple scattering by a set of particles is discussed at the end of this chapter, with the determination of the effective wave number of coherent waves propagating in heterogeneous media. For small concentration of particles, this effective wave number is close to that of the host medium and contains information on the scattering by a single particle.

In Chapter 3, we investigate the generation and detection of bulk and surface elastic waves. The most efficient and widespread technique for generating and detecting free bulk waves in a solid or a liquid, and also elastic guided waves in a plate or a cylinder, is to exploit the inverse and direct piezoelectric effects. The operating principle and the structure of transducers are described according to the frequency range. Their performances and their domain of application are specified. The properties of active materials commonly used (polarized piezoceramic or piezocomposite, piezoelectric polymer or single crystal) are given. In the second section, we analyze the generation and detection of surface acoustic waves, directly on a piezoelectric substrate, by transducers composed of two comb-shaped electrodes. Several models are developed to explain how these interdigital transducers operate. Their impulse and frequency responses, and their efficiency, are evaluated. The last two sections are devoted to optical methods, first the excitation of elastic waves by photothermal effects and second the detection of mechanical displacements by laser interferometry. The spatiotemporal characteristics of transmitted bulk and Rayleigh waves are given. These laser-based ultrasonic techniques present some advantages: they do not require any mechanical contact with the propagation medium; they can reach very high frequency domains and very small wavelengths. Therefore, they can be used for material evaluation at a large distance or with a high spatial resolution.

The authors would like to thank for their assistance in the preparation of this book:

– Claire Prada, Director of Research at the CNRS, at the Institute Langevin Ondes et Images;

– Jean-Marc Conoir, Director of Research at the CNRS, at the Institute Jean le Rond d'Alembert.

Tony VALIER-BRASIER
Daniel ROYER
February 2022

List of Main Symbols

a, b : amplitude of the waves.

c_{ijkl} (C_{IJ}) : stiffnesses (Voigt notation).

D : directivity factor.

$\underline{D}$: electric induction (displacement) vector.

E : Young modulus.

$\underline{E}$: electric field vector.

e_k (e_p) : kinetic (potential) energy per unit volume.

e_{ijk} (ε_{ij}) : piezoelectric (dielectric) constants.

$\underline{F}$: force density per unit mass.

$H(t)$: Heaviside step function.

I : acoustic intensity.

$\underline{J}$: Poynting vector.

K : bulk modulus.

$\underline{k}$ (k) : wave vector (number).

K_M : electromechanical coupling coefficient (mode M).

$\underline{l}$: unit vector normal to a surface.

$\underline{n}$: unit vector of the propagation direction.

P : transported power.

p_a : acoustic pressure.

$p\ (p_0)$: thermodynamic pressure (at rest).

$\underline{p}, \underline{q}$: polarization vectors.

$P_s\ (p_s)$: power supplied by the source (per unit volume).

$r\ (t)$: amplitude reflection (transmission) coefficients.

$R\ (T)$: intensity reflection (transmission) coefficients.

s : entropy per unit volume.

$\underline{s}$: phase slowness vector.

T : absolute temperature.

$\underline{T}$: mechanical traction, stress vector.

U : internal energy per unit volume.

$\underline{u}$: mechanical displacement vector.

$\underline{v}$: particle velocity vector.

$\underline{V}^{\mathrm{e}}$: energy velocity vector.

$V\ (V_g)$: phase (group) velocity.

$V_L\ (V_T)$: velocity of longitudinal (transverse) bulk waves.

$W\ (w)$: work (per unit volume).

$\underline{x}$: position vector.

X, Y, Z : crystallographic axes.

$Z_e\ (Y)$: electrical impedance (admittance).

$Z\ (\mathcal{Z})$: acoustic (mechanical) impedance.

α : attenuation coefficient.

$\underline{\underline{\Gamma}}\ (\underline{\gamma})$: elastic (piezoelectric) Christoffel tensor.

δ_{ij} : Kronecker's symbol.

$\delta(t)$: Dirac function.

$\underline{\underline{\varepsilon}}\ (\underline{\underline{S}})$: linearized strain tensor (piezoelectric solid).

η : viscosity coefficient.

$\theta_{i,r,t}$: angle of incidence, reflection, transmission.

Θ : dilatation.

κ : bulk wave velocity ratio (V_T/V_L).

λ, μ : Lamé constants.

ν : Poisson's ratio.

ρ : mass density.

$\rho_e\ (\sigma_e)$: electric charge density per unit volume (area).

$\sigma\ (\sigma_d)$: effective (differential) scattering cross-section.

$\underline{\underline{\sigma}}$: Cauchy stress tensor.

$\phi\ (\underline{\psi})$: scalar (vector) potential.

Φ : electric potential.

χ : decay factor.

ω : angular frequency.

1

Radiation of Elastic Waves

Acoustic waves emitted by sources of finite dimensions diverge, so that acoustic pressure and power density decrease with the distance from the transmitter. Understanding the radiation of sources is therefore essential to predict the acoustic field emitted by transducers used for medical diagnosis or in sonar systems, as well as in defect detectors in metallurgy or in other instruments and sensors.

The chapter is divided into three sections. In the first section, we study the acoustic radiation in a perfect fluid by a vibrating plane surface, then by a focused transducer. In the harmonic case and in a medium where the acoustic wave is defined by a scalar quantity such as the velocity potential or the acoustic pressure, the results are similar to those of the scalar theory of diffraction in optics. One specificity of acoustics is the use of short impulse waves or wave trains of varying durations. By drastically reducing the interferences between the waves emitted by the different points of the source, the conditions of the impulse diffraction, presented at the end of this section, are very different from those observed for time-harmonic waves. In the second section, we analyze the radiation of bulk elastic waves and Rayleigh waves in isotropic or anisotropic solids. With the exception of the explosive sources used in the field of oil exploration, these sources are most often located on the solid surface. The modeling of a seismic or a thermoelastic source is examined in the case of a linear distribution of impulsive forces. The last section is devoted to the radiation from an elementary spherical source embedded in an isotropic solid.

1.1. Acoustic radiation in a fluid

In a perfect fluid and assuming that the wave is a small perturbation of the reference state (pressure p_0, mass density ρ_0), the acoustic pressure $p_a = p - p_0$ and the particle velocity $\underline{v}$ are solutions to the local equation of dynamics, also called the linearized Euler equation:

$$\frac{\partial \underline{v}}{\partial t} + \frac{1}{\rho_0} \underline{\text{grad}}\ p_a = \underline{F}(\underline{x}, t), \qquad [1.1]$$

and to equation [1.322] from Volume 1:

$$\frac{\partial p_a}{\partial t} + \rho_0 c^2 \operatorname{div} \underline{v} = 0 \qquad [1.2]$$

resulting from the linearization of the continuity equation and the pressure–density relation of the fluid. $\underline{F}(\underline{x}, t)$ is the force density per unit mass and c is the sound speed. In order to solve this system of first-order partial differential equations, it is convenient to express the acoustic pressure and particle velocity in terms of a single scalar function, the velocity potential $\phi(\underline{x}, t)$, defined by:

$$\underline{v} = -\ \underline{\text{grad}}\ \phi \qquad \text{hence} \qquad \frac{\partial p_a}{\partial t} = \rho_0 c^2 \Delta\phi \qquad [1.3]$$

In the absence of sources in the volume ($\underline{F} = \underline{0}$), the acoustic pressure can be deduced from equation [1.1]:

$$p_a = \rho_0 \frac{\partial \phi}{\partial t} \qquad [1.4]$$

The time-derivation of equation [1.1] shows that the potential ϕ satisfies the equation:

$$\underline{\text{grad}} \left(c^2 \Delta\phi - \frac{\partial^2 \phi}{\partial t^2} \right) = \frac{\partial \underline{F}}{\partial t} \qquad [1.5]$$

By introducing the potential $s(\underline{x}, t)$ of the volume sources, defined by:

$$\frac{\partial \underline{F}}{\partial t} = -c^2\ \underline{\text{grad}}\ [s(\underline{x}, t)] \qquad [1.6]$$

equation [1.5] takes the form:

$$\Delta\phi - \frac{1}{c^2} \frac{\partial^2 \phi}{\partial t^2} = -s(\underline{x}, t) \qquad [1.7]$$

In the *harmonic case*, the velocity potential and volume source potential vary as $e^{-i\omega t}$:

$$\phi(\underline{x}, t) = \Phi(\underline{x}, \omega) e^{-i\omega t} \qquad \text{and} \qquad s(\underline{x}, t) = S(\underline{x}, \omega) e^{-i\omega t} \qquad [1.8]$$

The spectral components $\Phi(\underline{x}, \omega)$ and $S(\underline{x}, \omega)$, at angular frequency ω, satisfy the Helmholtz equation:

$$\Delta\Phi(\underline{x}, \omega) + k^2\Phi(\underline{x}, \omega) = -S(\underline{x}, \omega) \qquad [1.9]$$

where $k = \omega/c$ is the wave number.

1.1.1. *Integral formulation of the Helmholtz equation*

The determination of the velocity potential[1] $\Phi(\underline{x})$ in a closed domain is not always easy. One way to do this is based on the integral formulation of the Helmholtz equation, also called Kirchhoff–Helmholtz equation.

Let us consider a volume V delimited by a surface S of unit normal $\underline{n}$, oriented toward outside. The potential $\Phi(\underline{x})$, which is the solution to the Helmholtz equation [1.9], satisfies known boundary conditions on the surface S (Figure 1.1).

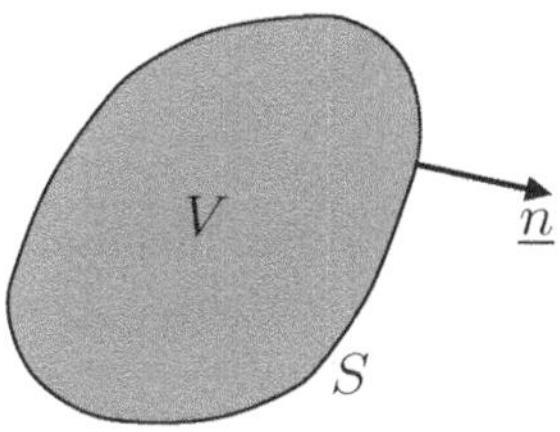

Figure 1.1. *Volume V delimited by the boundary S of outer unit normal $\underline{n}$*

Let $G(\underline{x}, \underline{x}_0)$ be the *Green's function*, solution to the Helmholtz equation in the domain V, whose source term is a Dirac delta function at $\underline{x}_0$:

$$\left(\Delta + k^2\right) G(\underline{x}, \underline{x}_0) = -\delta(\underline{x} - \underline{x}_0), \qquad \forall \underline{x}, \underline{x}_0 \in V \qquad [1.10]$$

The Green's function verifies the reciprocity principle, which is expressed by the relation:

$$G(\underline{x}, \underline{x}_0) = G(\underline{x}_0, \underline{x}) \qquad [1.11]$$

1 In the following and to simplify the notations, the reference to the angular frequency ω is removed for all quantities.

Multiplying equation [1.9] by $G(\underline{x}, \underline{x}_0)$, and equation [1.10] by $\Phi(\underline{x})$ and subtracting both results, we obtain:

$$G(\underline{x}, \underline{x}_0)\Delta\Phi(\underline{x}) - \Phi(\underline{x})\Delta G(\underline{x}, \underline{x}_0) = \delta(\underline{x} - \underline{x}_0)\Phi(\underline{x}) - S(\underline{x})G(\underline{x}, \underline{x}_0) \quad [1.12]$$

The integration of this equation over the volume V leads to the expression:

$$\Phi(\underline{x}_0) = \iiint_V [G(\underline{x}, \underline{x}_0)\Delta\Phi(\underline{x}) - \Phi(\underline{x})\Delta G(\underline{x}, \underline{x}_0)]\, \mathrm{d}^3\underline{x} + \iiint_V S(\underline{x})G(\underline{x}, \underline{x}_0)\, \mathrm{d}^3\underline{x} \quad [1.13]$$

or again, after using the second Green's identity:

$$\Phi(\underline{x}_0) = \iint_S \left[G(\underline{x}, \underline{x}_0)\frac{\partial\Phi(\underline{x})}{\partial n} - \Phi(\underline{x})\frac{\partial G(\underline{x}, \underline{x}_0)}{\partial n}\right] \mathrm{d}^2\underline{x} + \iiint_V S(\underline{x})G(\underline{x}, \underline{x}_0)\, \mathrm{d}^3\underline{x} \quad [1.14]$$

where $\partial\Phi(\underline{x})/\partial n$ is the derivative of the function $\Phi(\underline{x})$ in the direction of the outgoing normal. By switching the variables $\underline{x}$ and $\underline{x}_0$ and considering the reciprocity principle [1.11]:

$$\Phi(\underline{x}) = \iiint_V S(\underline{x}_0)G(\underline{x}, \underline{x}_0)\, \mathrm{d}^3\underline{x}_0 + \iint_S \left[G(\underline{x}, \underline{x}_0)\frac{\partial\Phi(\underline{x}_0)}{\partial n} - \Phi(\underline{x}_0)\frac{\partial G(\underline{x}, \underline{x}_0)}{\partial n}\right] \mathrm{d}^2\underline{x}_0 \quad [1.15]$$

the scalar potential $\Phi(\underline{x})$ is the sum of two integrals with different properties. The volume integral expresses the acoustic field generated by "primary" sources located inside the volume V. This integral is the convolution product of the Green's function $G(\underline{x}, \underline{x}_0)$ with the source function $S(\underline{x}_0)$. The surface integral expresses the effect of the boundary S of the domain on the acoustic field. This allows us to directly integrate the boundary conditions into the propagation equation, which can be useful in the case of complex surfaces, such as non-planar, irregular or vibrating surfaces. However, the analytical calculation of this integral can be difficult. Consequently, it is often carried out numerically.

1.1.2. *Rayleigh integral*

The objective of this section is to determine the field radiated by a plane surface S_0, whose points vibrate with the normal velocity $v_n(\underline{x}_0)e^{-i\omega t}$. The semi-infinite space $z > 0$, filled with a fluid, is bounded by the infinite plane S located at $z = 0$. This perfectly rigid plane S contains the emitting surface S_0. By hypothesis, the half-space $z > 0$ is free from volume source, so that the volume integral in equation [1.15] is zero. The velocity potential radiated by the surface S_0 of inner normal $\underline{e}_z$ is given by the surface integral[2]:

$$\Phi(\underline{x}) = \iint_S \left[\Phi(\underline{x}_0)\frac{\partial G(\underline{x},\underline{x}_0)}{\partial z_0} - G(\underline{x},\underline{x}_0)\frac{\partial \Phi(\underline{x}_0)}{\partial z_0}\right] \mathrm{d}S \qquad [1.16]$$

which expresses the Kirchhoff–Helmholtz theorem.

Given the relation [1.3] between the particle velocity and the potential Φ, the continuity of the normal velocity imposes:

$$v_n(\underline{x}_0) = -\frac{\partial \Phi}{\partial z_0} \qquad [1.17]$$

The velocity potential at point $\underline{x}$ takes the form:

$$\Phi(\underline{x}) = \iint_S \left[\Phi(\underline{x}_0)\frac{\partial G(\underline{x},\underline{x}_0)}{\partial z_0} + v_n(\underline{x}_0)G(\underline{x},\underline{x}_0)\right] \mathrm{d}S \qquad [1.18]$$

In the absence of any information on the value of the velocity potential on the surface S and in order to remove the term $\Phi(\underline{x}_0)$ in the integral equation, we must find a Green's function satisfying a Neumann condition ($\partial G/\partial z_0 = 0$) on this surface. The appropriate Green's function is obtained using the image-source method, by the summation of the free field Green's functions (Bruneau 2006):

$$G(\underline{x},\underline{x}_0) = \frac{e^{ik|\underline{x}-\underline{x}_0|}}{4\pi|\underline{x}-\underline{x}_0|} \qquad \forall \underline{x}_0 \qquad [1.19]$$

corresponding to a source located at point $\underline{x}_0$ and to the source image, with respect to the plane $z = 0$, located at $\underline{x}'_0$ (Figure 1.2):

$$G^+(\underline{x},\underline{x}_0) = \frac{e^{ik|\underline{x}-\underline{x}_0|}}{4\pi|\underline{x}-\underline{x}_0|} + \frac{e^{ik|\underline{x}-\underline{x}'_0|}}{4\pi|\underline{x}-\underline{x}'_0|} \qquad [1.20]$$

2 On the surface S, the outgoing unit normal is $-\underline{e}_z$, imposing a change of sign between equations [1.15] and [1.16].

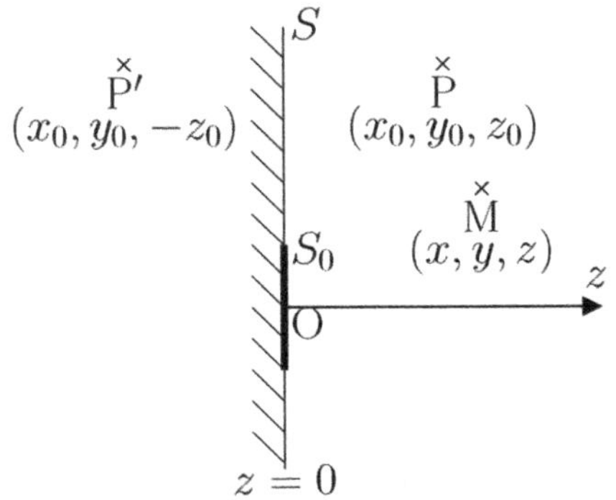

Figure 1.2. *Representation of the observation point M, the source P and the image source P'*

By writing:

$$\begin{cases} R = |\underline{x} - \underline{x}_0| = \sqrt{(x - x_0)^2 + (y - y_0)^2 + (z - z_0)^2} \\ R' = |\underline{x} - \underline{x}'_0| = \sqrt{(x - x_0)^2 + (y - y_0)^2 + (z + z_0)^2} \end{cases} \quad [1.21]$$

the derivative of the Green's function with respect to the normal to the surface $z = 0$ is given by:

$$\frac{\partial G^+(\underline{x}, \underline{x}_0)}{\partial z_0} = \frac{e^{ikR}}{4\pi R^2}(ikR - 1)\frac{\partial R}{\partial z_0} + \frac{e^{ikR'}}{4\pi (R')^2}(ikR' - 1)\frac{\partial R'}{\partial z_0} \quad [1.22]$$

or again, considering expressions [1.21] for R and R':

$$\frac{\partial G^+(\underline{x}, \underline{x}_0)}{\partial z_0} = \frac{e^{ikR}}{4\pi R^3}(ikR - 1)(z_0 - z) + \frac{e^{ikR'}}{4\pi (R')^3}(ikR' - 1)(z_0 + z) \quad [1.23]$$

At $z_0 = 0$, the distances R and R' are equal and therefore the normal derivative of the Green's function is zero on the surface $z_0 = 0$. The Green's function G^+ satisfies a Neumann condition on the surface S. Taking into account this condition and the expression [1.20] for a source point located at $z_0 = 0$:

$$G^+(\underline{x}, \underline{x}_0) = \frac{e^{ik|\underline{x} - \underline{x}_0|}}{2\pi |\underline{x} - \underline{x}_0|} \quad [1.24]$$

relation [1.18] can be written in the form of the *Rayleigh integral*:

$$\Phi(\underline{x}) = \frac{1}{2\pi} \iint_S v_n(\underline{x}_0) \frac{e^{ik|\underline{x} - \underline{x}_0|}}{|\underline{x} - \underline{x}_0|} \mathrm{d}S \qquad [1.25]$$

The Rayleigh integral was established in the harmonic case under the hypothesis that all points on the emitting surface S_0 vibrate in phase with a normal velocity $v_n(\underline{x}_0)e^{-i\omega t}$, *a priori* non-uniform. In the *piston mode*, the amplitude distribution can be weighted by a function $P_n(\underline{x}_0)$, which is smooth enough to reduce the edge effects and thus the level of the side lobes. In the *uniform piston* mode, all the points on the source have the same velocity $v_0(t)$; the function $P_n(\underline{x}_0)$ is constant and chosen to be equal to unity on the surface S_0, and it is zero outside.

This distinction is independent of the type of excitation, that is, of the operating regime (harmonic, impulse, transient, etc.), as shown in Table 1.1.

Mode	Normal velocity	Velocity potential
Harmonic	$v_n = P_n(\underline{x}_0)v_0e^{-i\omega t}$	$\Phi(\underline{x}, \omega) = v_0 H(\underline{x}, \omega)$
Impulse	$v_n = P_n(\underline{x}_0)v_0\delta(t)$	$\phi(\underline{x}, t) = v_0 h(\underline{x}, t)$
Transient	$v_n = P_n(\underline{x}_0)v_0(t)$	$\phi(\underline{x}, t) = v_0(t) \otimes h(\underline{x}, t)$

Table 1.1. *Piston mode: main operating regimes*

REMARK.– The velocity potential $\Phi(\underline{x}, \omega)$ is naturally the Fourier transform of $\phi(\underline{x}, t)$. It is the same for the frequency response $H(\underline{x}, \omega)$ and the impulse response $h(\underline{x}, t)$.

There exists another operating mode specific to multi-element antennas (or arrays) used in sonars or ultrasound scanners. The source consists of a large number of piezoelectric components, usually rectangular, whose width is small compared to the wavelength. The electric voltage applied to each element is either delayed (in the impulse regime) or phase-shifted (in the harmonic case) according to a linear law, depending on the position of the elements (section 1.1.4.4). The electronic control of the delay or of the phase-shift law ensures the dynamic scanning of these ultrasound imaging devices.

1.1.3. *Rayleigh–Sommerfeld integral*

If we choose the difference of the free field Green's functions, the Green's function obtained:

$$G^{-}(\underline{x}, \underline{x}_0) = \frac{e^{ikR}}{4\pi R} - \frac{e^{ikR'}}{4\pi R'} \qquad [1.26]$$

is zero at any point on the plane surface S, because $R = R'$ when $z_0 = 0$. By changing the + sign to the – sign in the right-hand side of equation [1.23], the normal derivative to the surface S is expressed as:

$$\left.\frac{\partial G^{-}}{\partial z_0}\right|_{z_0=0} = -\frac{e^{ikR}}{2\pi R^3}(ikR - 1)z \qquad [1.27]$$

Let us carry out two approximations, justified in optics, to calculate the field diffracted by an aperture:

– the incident wave is little modified on the surface of the aperture, whose dimensions are large with respect to the wavelength λ;

– the observation point M is located at a large distance, that is, $kR >> 1$.

Under these conditions, equation [1.18] takes the form:

$$\phi_d(\underline{x}) = \iint_S \phi_i(\underline{x}_0)\frac{e^{ikR}}{R}\frac{\cos\theta}{i\lambda}\,\mathrm{d}S \qquad \text{with} \qquad R = |\underline{x} - \underline{x}_0| \qquad [1.28]$$

where θ is the angle between the vector $\underline{x} - \underline{x}_0$ and the z axis ($\cos\theta = z/R$). This *Rayleigh–Sommerfeld formula* provides the field $\phi_d(\underline{x})$ diffracted by a plane aperture over which the incident field $\phi_i(\underline{x}_0)$ is known. This mathematically expresses the Huygens–Fresnel principle, by giving a value $\cos\theta/i\lambda$ to the complex amplitude of the hemispherical wavelets emitted, in the propagation direction of the incident wave, at each point on the diffracting surface. The comparison between the Rayleigh integral and the Rayleigh–Sommerfeld integral explains the similarity between the acoustic radiation of a uniform piston [$v_n(\underline{x}_0) = $ Const.] at a large distance ($R >> \lambda$) and the diffraction of a plane light wave [$\phi_i(\underline{x}_0) = $ Const.] through an aperture whose shape is identical to that of the piston.

1.1.4. *Harmonic case*

We start by analyzing the radiation of time-harmonic acoustic waves by planar sources such as a circular disk, a rectangular element, or a multi-element linear antenna. These reference sources serve as a model for piezoelectric transducers used in various fields, such as non-destructive evaluation of materials, medical echography and underwater exploration by ultrasound.

1.1.4.1. *Plane circular piston*

The case of a plane baffled piston is a major problem in acoustics, as it allows us to model many acoustic sources, such as loudspeakers or piezoelectric transducers. The circular source of radius a, vibrating uniformly, is included in an infinite rigid screen (Figure 1.3). The calculation of the Rayleigh integral requires the use of polar and spherical coordinates. The position of a source point P on the surface of the disk is defined by the vector $\underline{x}_0 = (r_0 \cos\varphi_0, r_0 \sin\varphi_0, 0)$ and the observation point M in the domain V is marked by the vector $\underline{x} = (r\cos\varphi \sin\theta, r\sin\varphi \sin\theta, r\cos\theta)$.

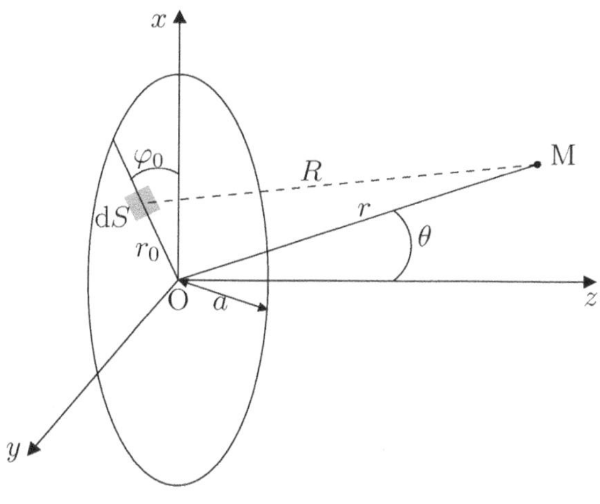

Figure 1.3. *The particle velocity on the disk is uniform (constant amplitude v_0). The observation point M is assumed to be in the xz plane*

The particle velocity, normal to the plane surface, is assumed to be uniform:

$$v_n(\underline{x}_0) = \begin{cases} v_0 & \text{if } |\underline{x}_0| \leq a \\ 0 & \text{if } |\underline{x}_0| > a \end{cases} \qquad [1.29]$$

Since the problem is axisymmetric, calculations can be performed by taking $\varphi = 0$, without any loss of generality. The distance $R = |\underline{x} - \underline{x}_0|$ between the source point and the observation point is equal to:

$$|\underline{x} - \underline{x}_0| = \sqrt{(r\sin\theta - r_0\cos\varphi_0)^2 + r_0^2\sin^2\varphi_0 + r^2\cos^2\theta} \\ = \sqrt{r^2 + r_0^2 - 2rr_0\sin\theta\cos\varphi_0} \quad [1.30]$$

and the potential at the observation point M is given by:

$$\Phi(r,\theta) = \frac{v_0}{2\pi}\int_0^a\int_0^{2\pi}\frac{e^{ik\sqrt{r^2+r_0^2-2rr_0\sin\theta\cos\varphi_0}}}{\sqrt{r^2+r_0^2-2rr_0\sin\theta\cos\varphi_0}}r_0\,\mathrm{d}r_0\,\mathrm{d}\varphi_0 \quad [1.31]$$

This integral cannot be expressed analytically in the general case, but it is possible to evaluate it exactly along the piston axis and approximately in the far field.

1.1.4.2. *Radiated field on the piston axis*

In the harmonic case, the acoustic pressure $P_a(\underline{x},\omega)$ is related to the velocity potential through relation [1.4]:

$$P_a(\underline{x}) = -i\omega\rho_0\Phi(\underline{x}) = -ikZ_0\Phi(\underline{x}) \quad [1.32]$$

where $Z_0 = \rho_0 c$ is the acoustic impedance of the fluid. On the piston axis ($\theta = 0$), this pressure can be expressed as:

$$P_a(z,\theta=0) = -ikZ_0\frac{v_0}{2\pi}\int_0^a\int_0^{2\pi}\frac{e^{ikR}}{R}r_0\,\mathrm{d}r_0\,\mathrm{d}\varphi_0 \quad [1.33]$$

By performing the change of variable $R^2 = z^2 + r_0^2$, the acoustic pressure takes the form:

$$P_a(z,\theta=0) = -ikZ_0v_0\int_z^{\sqrt{z^2+a^2}} e^{ikR}\,\mathrm{d}R = Z_0v_0\left[e^{ikz} - e^{ik\sqrt{z^2+a^2}}\right] \quad [1.34]$$

Thus, the pressure field results from the interference of a traveling plane wave emitted by the whole piston, and a wave generated by diffraction on the edge of the piston.

The amplitude of the pressure field normalized by Z_0v_0 is plotted in Figure 1.4, as a function of the ratio z/a for four different values of the product ka: (a) $ka = 1$, (b) $ka = 10$, (c) $ka = 50$ and (d) $ka = 100$. For $ka = 1$, that is, when the wavelength $\lambda = 2\pi a$ is larger than the piston radius, the acoustic pressure on the piston axis decreases rapidly. If the wavelength is comparable to the radius ($ka = 10$), this pressure passes through a single maximum and then decreases as z increases. In situations where the radius a is greater than the wavelength λ, the amplitude of the normalized pressure oscillates between 0 and 2 and then subsequently decreases. The larger the product ka is, the greater the number of oscillations in the near field. The position of the last maximum marks the limit between the *Fresnel zone*, representing the near field, and the *Fraunhofer zone*, representing the far field.

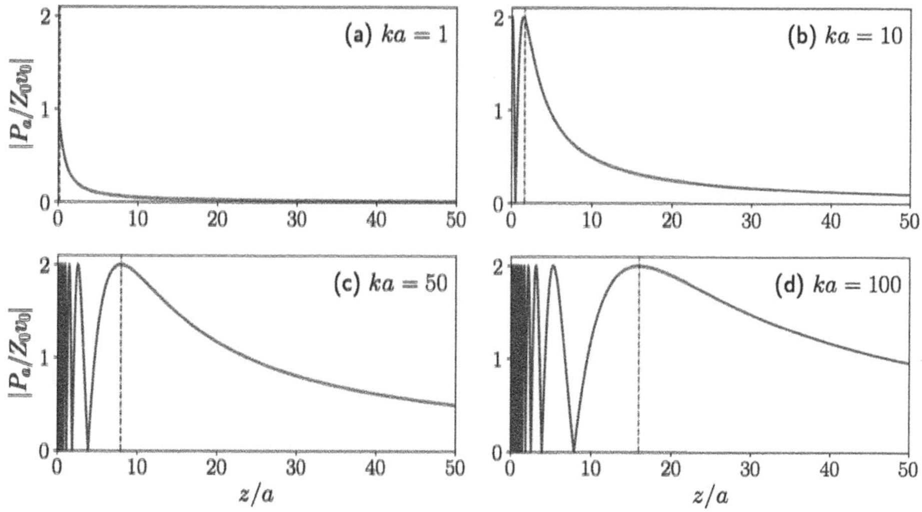

Figure 1.4. *Amplitude of the normalized pressure field radiated by a uniform piston of radius a along its axis for four values of the product ka*

1.1.4.2.1. Near field: Fresnel zone

The positions of the pressure maxima and zeros are easily deduced from its modulus. According to relation [1.34]:

$$|P_a(z,\theta=0)|^2 = 4(Z_0 v_0)^2 \sin^2\left[\frac{k}{2}\left(\sqrt{z^2+a^2}-z\right)\right] \quad [1.35]$$

the expression for the modulus of the pressure normalized by $Z_0 v_0$ is:

$$\frac{|P_a(z,\theta=0)|}{Z_0 v_0} = 2\left|\sin\left[\frac{k}{2}\left(\sqrt{z^2+a^2}-z\right)\right]\right| \quad [1.36]$$

The pressure maxima result from *constructive interference*. They correspond to the extreme values of the sine function:

$$\frac{k}{2}\left(\sqrt{z_m^2+a^2}-z_m\right)=m\frac{\pi}{2} \qquad \text{with} \qquad m=2n+1 \tag{1.37}$$

therefore, for the values of z that satisfy relation:

$$z_m=\frac{(ka)^2-m^2\pi^2}{2mk\pi} \qquad \text{with} \qquad m=2n+1 \tag{1.38}$$

The maximum that is farthest from the source is located at the distance z_1 ($m=1$):

$$z_1=\frac{(ka)^2-\pi^2}{2k\pi} \tag{1.39}$$

The pressure cancellations result from *destructive interference*, which takes place when the argument of the sine is equal to $n\pi$, that is, for values of z_n that satisfy relation:

$$z_n=\frac{(ka)^2-(2n\pi)^2}{4nk\pi} \qquad \text{with} \qquad n\in\mathbb{N} \tag{1.40}$$

1.1.4.2.2. Far field: Fraunhofer zone

The position of the last pressure maximum is often used as the lower limit of the far field. If the piston radiates at high frequency, the product ka is large and the position of the last pressure maximum is:

$$z_1\approx\frac{ka^2}{2\pi}=\frac{a^2}{\lambda} \tag{1.41}$$

This expression remains valid as long as the distance z between the far field input and the source is large compared to the source diameter and to the wavelength.

ORDER OF MAGNITUDE.– In immersion experiments, it may be important to place the sample to be probed in the far field of the emitter. For a transducer of diameter $2a=13$ mm, which generates longitudinal waves in water at a frequency $f=10$ MHz, the far field corresponds to $z>a^2/\lambda\approx 28$ cm.

In the far field, the distance z is large in front of the radius a. A limited development over the quantity $\sqrt{z^2+a^2}$ shows that the pressure decreases as $1/z$:

$$P_a(z,\theta=0)\approx Z_0v_0e^{ikz}\left(1-e^{ik\frac{a^2}{2z}}\right)\approx -ika^2Z_0v_0\frac{e^{ikz}}{2z} \tag{1.42}$$

1.1.4.3. *Directivity in the far field*

As previously mentioned, the Fraunhofer zone corresponds to the far field, that is, to distances r larger than the radius a of the piston. In this case, the distance $|\underline{x} - \underline{x}_0|$ appearing in the expression [1.31] of the velocity potential can be approximated by:

$$\begin{aligned} |\underline{x} - \underline{x}_0| &= r\sqrt{1 + (r_0/r)^2 - 2(r_0/r)\sin\theta\cos\varphi_0} \\ &\approx r\,[1 - (r_0/r)\sin\theta\cos\varphi_0] \end{aligned} \qquad [1.43]$$

so that:

$$\Phi(r,\theta) = \frac{v_0}{2\pi}\int_0^a\int_0^{2\pi}\frac{e^{ikr[1-(r_0/r)\sin\theta\cos\varphi_0]}}{r\,[1 - (r_0/r)\sin\theta\cos\varphi_0]}r_0\,\mathrm{d}r_0\,\mathrm{d}\varphi_0 \qquad [1.44]$$

or again:

$$\Phi(r,\theta) \approx v_0\frac{e^{ikr}}{r}\int_0^a r_0\int_0^{2\pi}\frac{e^{-ikr_0\sin\theta\cos\varphi_0}}{2\pi}\,\mathrm{d}\varphi_0\,\mathrm{d}r_0 \qquad [1.45]$$

Knowing that the zero-order and first-order cylindrical Bessel functions satisfy the relations:

$$J_0(x) = \frac{1}{2\pi}\int_0^{2\pi} e^{ix\cos\alpha}\,\mathrm{d}\alpha \qquad \text{and} \qquad \int xJ_0(x)\,\mathrm{d}x = xJ_1(x) \qquad [1.46]$$

with $x = kr_0\sin\theta$, the velocity potential takes the form:

$$\Phi(r,\theta) = v_0\frac{e^{ikr}}{(k\sin\theta)^2 r}\int_0^{ka\sin\theta} xJ_0(x)\,\mathrm{d}x = Q\frac{e^{ikr}}{2\pi r}D(\theta) \qquad [1.47]$$

$Q = \pi a^2 v_0$ is the piston flow and $D(\theta)$ is the directivity factor, given by:

$$D(\theta) = \frac{2J_1(ka\sin\theta)}{ka\sin\theta} \qquad [1.48]$$

Since $J_1(x) = x/2$ for small values of $x = ka\sin\theta$, the function $D(\theta)$ is maximal and equal to unity for $\theta = 0$.

In Figure 1.5, the directivity factor is plotted for two values of the product ka. If ka is small compared to unity, that is, at low frequency, the directivity factor is close to unity and depends little on θ. The radiation is omnidirectional and the source behaves like a point source (which is in agreement with the fact that its dimensions are small compared to the wavelength). When the product ka increases, the number of side lobes increases and the width of the main lobe shrinks: the piston is more directional at high frequency.

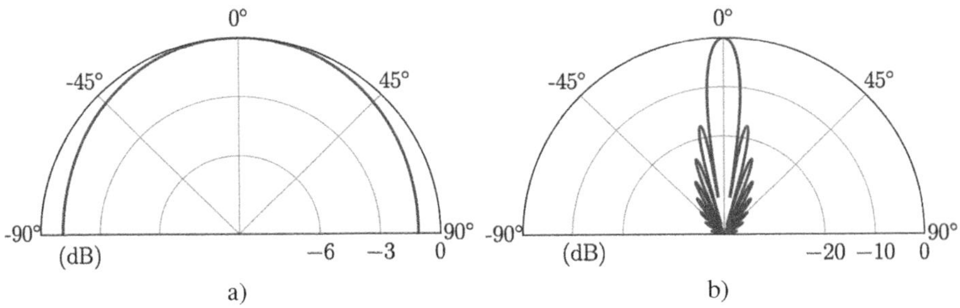

Figure 1.5. *Directivity patterns of a circular piston of radius a for a) $ka = 1$ and b) $ka = 25$. The level of the first side lobe is 17.5 dB below that of the main lobe*

Two parameters are used to characterize the directivity of the emitted acoustic beam:

– *The divergence angle* defined by the angular difference $2\theta_0$ between the zeros on either side of the main lobe. Knowing that the first zero of the Bessel function $J_1(x)$ corresponds to the argument $x = 3.83$, the divergence angle of the beam is given by:

$$2\theta_0 \approx 2\arcsin\left(\frac{3.83}{ka}\right) = 2\arcsin\left(0.61\frac{\lambda}{a}\right) \qquad [1.49]$$

Since the maximum of the first side lobe is equal to only 0.133, that is –17 dB relative to that of the central lobe, the energy emitted by the disk is contained, for the most part, in the cone of apex angle $2\theta_0 \approx 1.22\lambda/a$, if $\lambda << a$.

– *The –3 dB angular aperture* is defined by the angular difference $2\theta_{1/2}$ between the two directions (in a meridian plane) for which the acoustic power is divided by two. It is twice the angle $\theta_{1/2}$ for which the directivity factor is divided by $\sqrt{2}$, that is, $D(\theta_{1/2}) = 1/\sqrt{2}$:

$$2\theta_{1/2} \approx 2\arcsin\left(\frac{1.62}{ka}\right) = 2\arcsin\left(0.26\frac{\lambda}{a}\right) \qquad [1.50]$$

The emission by the piston disk is more directional when its diameter $2a$ is large with respect to the wavelength λ.

1.1.4.4. *Linear array*

The antennas (transmitters and receivers) of underwater exploration and localization systems (sonar) or the arrays of medical ultrasound equipment are made up of a large number of rectangular piezoelectric elements. The direction of the emitted acoustic beam is controlled by applying a delay law or a phase law to the different elements. The angular width of the swept sector is limited by the aperture of the beam emitted by an isolated element, whose width must be smaller than the wavelength λ. The angular resolution is all the better as the length L of the antenna is large compared to λ. Therefore, the number $2N$ of elements of the antenna must be large, typically $32 \leq 2N \leq 256$.

1.1.4.4.1. Rectangular element

Let us first examine the acoustic radiation in the far field of a rectangular element of width $2d$ (along x) and height $2h$ (along y), located at $z = 0$ and included in the surface S (Figure 1.6). A point on its surface S_0 is defined by the vector $\underline{x}_0 = (x_0, y_0, 0)$, where $x_0 \in [-d, d]$ and $y_0 \in [-h, h]$. An observation point M is marked in the domain V by the vector $\underline{x} = (r \cos\varphi \sin\theta, r \sin\varphi \sin\theta, r \cos\theta)$.

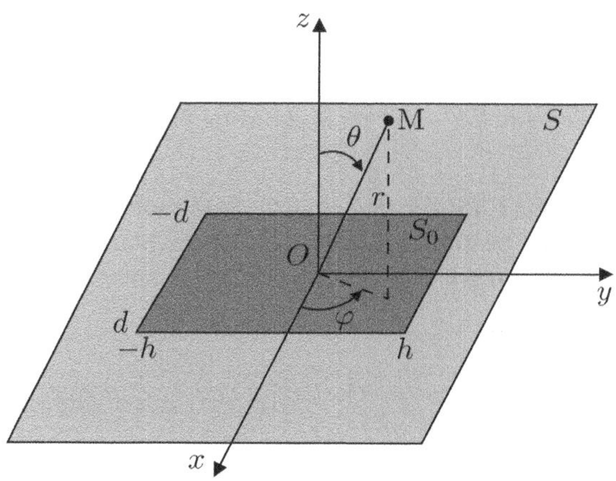

Figure 1.6. *Uniform rectangular element of width $2d$, height $2h$ and area $S_0 = 4dh$. The field is calculated at the position $\underline{x} = \underline{OM}$*

Assuming that x_0 and y_0 are small compared to r, the norm of vector $|\underline{x} - \underline{x}_0|$ is expressed by:

$$|\underline{x} - \underline{x}_0| \approx r - x_0 \cos\varphi \sin\theta - y_0 \sin\varphi \sin\theta \qquad [1.51]$$

In the harmonic case, the velocity potential results from the Rayleigh integral [1.25]; in the far field, its approximate expression becomes:

$$\Phi(\underline{x}) = \frac{e^{ikr}}{2\pi r}\int_{-d}^{d}\int_{-h}^{h} v_n(x_0, y_0)e^{-ik(x_0\cos\varphi\sin\theta + y_0\sin\varphi\sin\theta)}\,dx_0\,dy_0 \quad [1.52]$$

Since the vibratory velocity v_n is zero in the plane S, removed from S_0, the bounds of the two integrals can be extended to infinity. The velocity potential is then given by

$$\Phi(r,\theta,\varphi) = \frac{e^{ikr}}{2\pi r}\hat{V}_n(k\cos\varphi\sin\theta, k\sin\varphi\sin\theta) \quad [1.53]$$

where $\hat{V}_n$ is the double spatial Fourier transform, defined by the relation:

$$\hat{V}_n(k_x, k_y) = \int_{-\infty}^{\infty}\int_{-\infty}^{\infty} v_n(x_0, y_0)e^{-i(k_x x_0 + k_y y_0)}\,dx_0\,dy_0 \quad [1.54]$$

This formulation is general: in the far field, the velocity potential emitted by a plane surface S vibrating in the piston mode ($v_n = v_0 P_n(\underline{x}_0)$) is proportional to the spatial Fourier transform of the amplitude distribution $P_n(\underline{x}_0)$ of the normal velocity:

$$\Phi(\underline{x}) = \frac{v_0}{2\pi r}e^{ikr}\hat{P}_n(\underline{k}) \qquad \text{with} \qquad \hat{P}_n(\underline{k}) = \iint_S P_n(\underline{x}_0)e^{-i\underline{k}\cdot\underline{x}_0}\,dS \quad [1.55]$$

Thus, the expression [1.45] for the acoustic field radiated by the uniform circular piston in Figure 1.3 can be found directly. In the xz plane ($\varphi = 0$), equivalent to any plane due to the symmetry of revolution: $k_x = k\sin\theta$, $k_y = 0$, we obtain with $x_0 = r_0\cos\varphi_0$:

$$\hat{P}_n(\theta) = \int_0^a r_0 \int_0^{2\pi} e^{-ikr_0\sin\theta\cos\varphi_0}\,d\varphi_0\,dr_0 = \pi a^2 D(\theta) \quad [1.56]$$

where $D(\theta)$ is the directivity factor [1.48].

If the rectangular element vibrates in a uniform piston mode, the normal velocity is independent of the point:

$$v_n(x_0, y_0) = \begin{cases} v_0 & \text{if } (x_0, y_0) \in S_0 \\ 0 & \text{otherwise} \end{cases} \quad [1.57]$$

The expression for the velocity potential is:

$$\Phi(r,\theta,\varphi) = Q\frac{e^{ikr}}{2\pi r}D(\theta,\varphi) \quad [1.58]$$

where $Q = 4dhv_0$ is the piston flow and the directivity factor $D(\theta, \varphi)$ is a double cardinal sine function:

$$D(\theta, \varphi) = \text{sinc}\,(kd \cos\varphi \sin\theta)\,\text{sinc}\,(kh \sin\varphi \sin\theta) \qquad [1.59]$$

The radiation in the far field of a rectangular element of width $2d = 1$ mm and height $2h = 3$ mm vibrating at 10 MHz is represented in Figure 1.7. The field is calculated at a distance $r = 50$ mm. The color scale is saturated in order to visualize the side lobes.

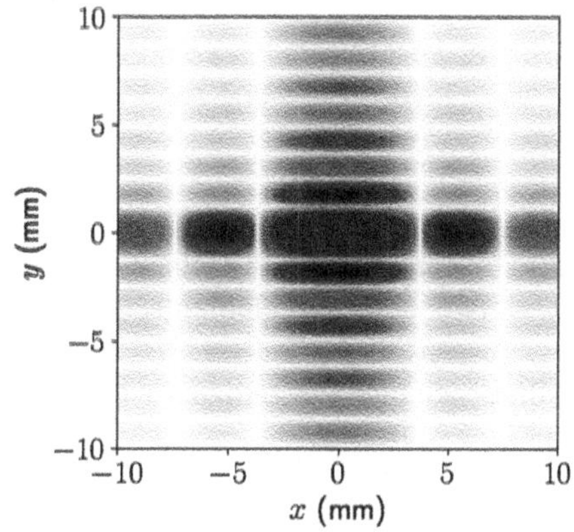

Figure 1.7. *Radiation in the far field ($r = 50$ mm) of a plane rectangular piston, of dimensions $2d = 1$ mm and $2h = 3$ mm, at a frequency $f = 10$ MHz. For a color version of this figure, see www.iste.co.uk/royer/waves2.zip*

1.1.4.4.2. Linear antenna

Let us now consider the case of a distribution of $2N$ linear rectangular elements of width $2d$ and height $2h$ (Figure 1.8). The velocity potential is given by the relation:

$$\Phi(\underline{x}) = \frac{1}{2\pi} \int_{-h}^{h} \int_{-d}^{d} \sum_{m=0}^{N-1} \left[v_m^-(x_0, y_0) \frac{e^{ikR_m^-}}{R_m^-} + v_m^+(x_0, y_0) \frac{e^{ikR_m^+}}{R_m^+} \right] \mathrm{d}x_0 \, \mathrm{d}y_0 \qquad [1.60]$$

with, in the far field:

$$\begin{cases} R_m^- \approx r + [(2m+1)d - x_0] \cos\varphi \sin\theta - y_0 \sin\varphi \sin\theta \\ R_m^+ \approx r - [(2m+1)d + x_0] \cos\varphi \sin\theta - y_0 \sin\varphi \sin\theta \end{cases} \qquad [1.61]$$

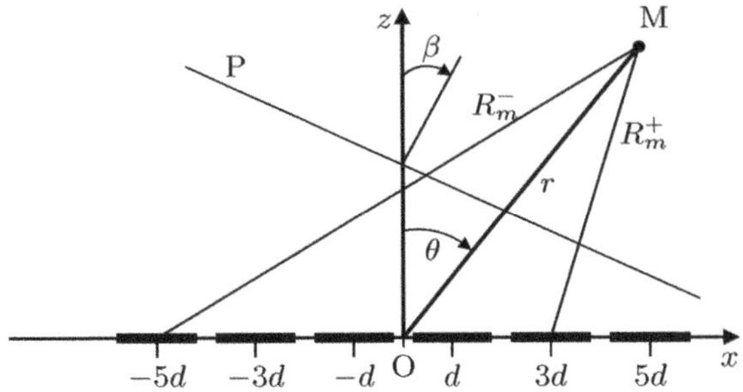

Figure 1.8. *Linear antenna composed of* $2N$ *sources aligned along the* x *axis*

In order for the plane P, inclined at an angle β with respect to the normal, to be equiphase, it is necessary to compensate the phase difference $\Delta\psi = 2kd\sin\beta$ between waves emitted by two adjacent elements, $2d$ apart, by applying to the emission a phase law $\psi_m = \psi_0 - kx_m\sin\beta$, linear *versus* the abscissa $x_m = \pm(2m+1)d$. Assuming that each element radiates with the same flow Q and with a uniform velocity equal to:

$$v_m^{\pm} = \frac{Q}{4dh}e^{\pm i\psi_m} \qquad \text{with} \qquad \psi_m = (2m+1)kd\sin\beta \qquad [1.62]$$

the velocity potential in the xz plane ($\varphi = 0$) takes the form:

$$\Phi(\underline{x}) = \frac{Qe^{ikr}}{4d\pi r}\sum_{m=0}^{N-1}\left[e^{i(2m+1)\psi} + e^{-i(2m+1)\psi}\right]\int_{-d}^{d} e^{-ikx_0\sin\theta}\,\mathrm{d}x_0 \qquad [1.63]$$

with $\psi = kd(\sin\theta - \sin\beta)$. Given expression [1.59] for the directivity factor of a rectangular element:

$$\Phi(\underline{x}) = \frac{Qe^{ikr}}{2\pi r}D(\theta,\varphi=0)\sum_{m=0}^{N-1}\left[e^{i(2m+1)\psi} + e^{-i(2m+1)\psi}\right] \qquad [1.64]$$

and by recognizing the geometric series $\sum_m q^m$, we obtain:

$$\Phi(\underline{x}) = \frac{Qe^{ikr}}{2\pi r}D(\theta,\varphi=0)\frac{\sin\left[2Nkd(\sin\theta - \sin\beta)\right]}{\sin\left[kd(\sin\theta - \sin\beta)\right]} \qquad [1.65]$$

In Figure 1.9, the radiation pattern for a linear antenna composed of $2N = 64$ rectangular elements, of width $2d = 4$ mm and height $2h = 4$ cm, is plotted in the

xz plane ($\varphi = 0$) as a function of angle θ for three phase-shift angles: $\beta = 0°$, 30° and 60°. This example is representative of an equipment for shallow-water sonar (depth 5–500 m) operating at 100 kHz ($\lambda = 15$ mm). The width of each element, of the order of $\lambda/4$, is small enough to ensure an efficient scanning over a 100° sector. For $\beta > 50°$, the emitted amplitude decreases due to the elemental directivity factor $D(\theta, \varphi)$, and the angular aperture of the beam increases. This characteristic is defined by the angle $\Delta\theta$ between two directions in the xz plane that are symmetric with respect to the beam axis and for which the emitted power is divided by two. For an antenna of length $L = 4Nd$ and a small angle θ, the directivity function [1.65] takes the form:

$$D_{ant} \approx 2N\frac{\sin\alpha}{\alpha} \qquad \text{with} \qquad \alpha = \pi\frac{L}{\lambda}\sin\theta \qquad [1.66]$$

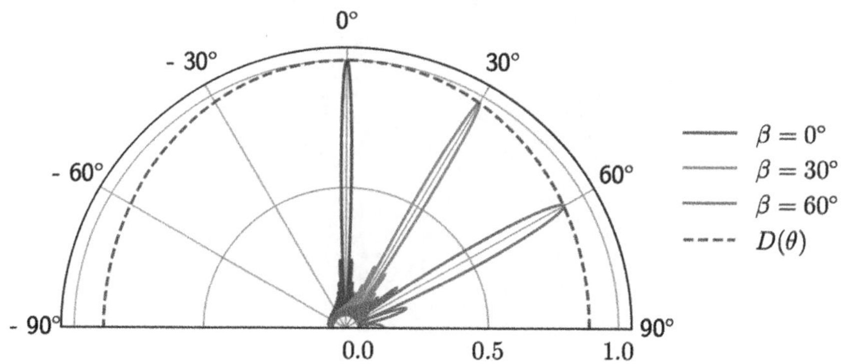

Figure 1.9. *Radiation patterns of a linear antenna composed of 64 elements of width $2d = 4$ mm and height $2h = 4$ cm at a frequency $f = 100$ kHz. For a color version of this figure, see www.iste.co.uk/royer/waves2.zip*

Its maximum value ($2N$) is divided by $\sqrt{2}$ when $\alpha = \pm 0.885\pi/2$, that is:

$$\sin\theta \approx \pm 0.885\frac{\lambda}{2L} \qquad \longrightarrow \qquad \Delta\theta = 0.885\frac{\lambda}{L}\,(\text{rad}) \approx 50\frac{\lambda}{L}\,(\text{deg}) \qquad [1.67]$$

i.e. $\Delta\theta \approx 3°$ for $L = 256$ mm and $\lambda = 15$ mm.

Figure 1.10 is an example of the measurement of the dispersion curves of a 1-mm thick duralumin plate using a linear array composed of 128 rectangular elements of dimensions 0.2 $\times 10$ mm^2 and spaced apart by 0.25 mm. A film of water ensures the coupling between the probe and the sample. A programmable electronic device transmits pulses to the first element, of center frequency 8 MHz and duration 4 μs, whose spectrum ranges from 4 to 12 MHz. This electronic device records the

128 signals received by all the elements. The Fourier transforms with respect to the time and space of these signals, acquired over 25 μs, are stored in a matrix. The N non-zero components are stored in a $N \times 3$ (k_n, f_n, w_n) matrix, where k, f and w represent, respectively, the wave number, frequency and Fourier coefficient. The representation in the "frequency-wave number" plane reveals the Lamb modes, characteristic of the propagation in a plate (section 4.3.1.2 in Volume 1). As noted in section 4.3.2 of Volume 1, the coupling with water does little modify the phase velocity of Lamb modes (Bochud *et al.* 2018).

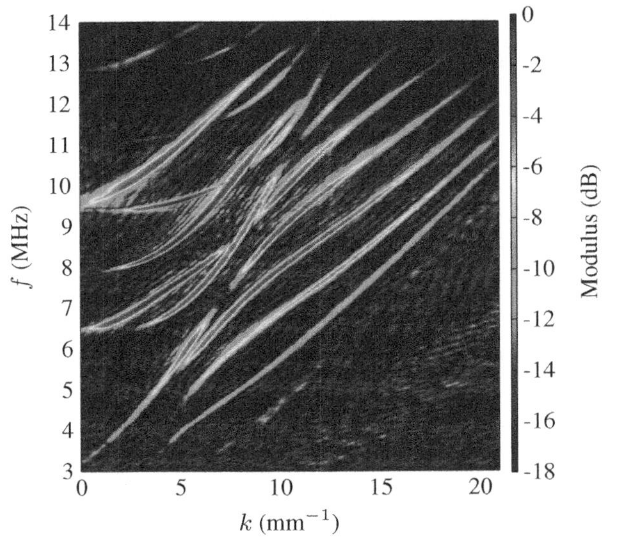

Figure 1.10. *Experimental dispersion curves of Lamb waves propagating in a duralumin plate of thickness 1 mm, measured using an array of 128 rectangular elements. For a color version of this figure, see www.iste.co.uk/royer/waves2.zip*

1.1.5. *Impulse diffraction*

In domains like medical ultrasound imaging or non-destructive testing of materials, transducers that generate acoustic waves are often excited by short pulses. The velocity potential at point M and angular frequency ω is obtained by replacing the wave number k by ω/c in the expression [1.25] for the Rayleigh integral, established in the harmonic case:

$$\Phi(\underline{x}, \omega) = \iint_S V_n(\underline{x}_0, \omega) \frac{e^{i\omega R/c}}{2\pi R} \, dS \qquad \text{with} \qquad R = |\underline{x} - \underline{x}_0| \qquad [1.68]$$

$\Phi(\underline{x}, \omega)$ is the spectral component, at angular frequency ω, of the instantaneous potential $\phi(\underline{x}, t)$ at the observation point M. Similarly, $V_n(\underline{x}_0, \omega)$ is the spectral component of the normal velocity $v_n(\underline{x}_0, t)$ at point P on the emitting surface. The inverse Fourier transform:

$$\phi(\underline{x}, t) = \frac{1}{2\pi} \int_{-\infty}^{+\infty} \Phi(\underline{x}, \omega) e^{-i\omega t} \, d\omega \qquad [1.69]$$

is written as given below, by changing the integration order:

$$\phi(\underline{x}, t) = \iint_S \frac{1}{2\pi R} \left[\int_{-\infty}^{+\infty} V_n(\underline{x}_0, \omega) \frac{e^{-i\omega(t - R/c)}}{2\pi} \, d\omega \right] dS \qquad [1.70]$$

The integral in square brackets represents the normal component of the particle velocity at point P and time $t - R/c$, that is, $v_n(\underline{x}_0, t - R/c)$. Thus, *the Rayleigh integral* can be expressed in the time domain by:

$$\phi(\underline{x}, t) = \iint_S \frac{v_n(\underline{x}_0, t - R/c)}{2\pi R} \, dS \qquad [1.71]$$

1.1.5.1. *Piston mode*

Let us consider the case where all points of the source vibrate with the same normal velocity $v_0(t)$:

$$v_n(\underline{x}_0, t) = P_n(\underline{x}_0) v_0(t) \qquad [1.72]$$

where $P_n(\underline{x}_0)$ is the spatial amplitude profile of the vibration over the piston. Since the Dirac delta function $\delta(t)$ is the neutral element of the convolution operator, the function $v_0(t - R/c)$ can be written as:

$$v_0 \left(t - \frac{R}{c} \right) = v_0(t) \otimes \delta \left(t - \frac{R}{c} \right) \qquad [1.73]$$

By substituting equations [1.72] and [1.73] into relation [1.71], we get:

$$\phi(\underline{x}, t) = v_0(t) \otimes \iint_S \frac{P_n(\underline{x}_0)}{2\pi R} \delta \left(t - \frac{R}{c} \right) dS = v_0(t) \otimes h(\underline{x}, t) \qquad [1.74]$$

with:

$$h(\underline{x}, t) = \iint_S \frac{P_n(\underline{x}_0)}{2\pi R} \delta \left(t - \frac{R}{c} \right) dS \qquad [1.75]$$

The velocity potential at point M is the convolution product of the velocity $v_0(t)$ at the piston surface with the space-time function $h(\underline{x}\,,t)$, called the *diffraction impulse response* at point M and time t. This result expresses the linearity and time invariance of the model (Fink and Cardoso 1984). This concept of an impulse response, that is, the response to a pulse shorter than the smallest characteristic time of the system, is commonly used in signal processing.

The Fourier transform of a convolution product is the simple product of transforms:

$$\Phi(\underline{x}\,,\omega) = H(\underline{x}\,,\omega)V_0(\omega) \qquad [1.76]$$

the *diffraction frequency response* $H(\underline{x}\,,\omega)$ being the Fourier transform of $h(\underline{x}\,,t)$:

$$H(\underline{x}\,,\omega) = \iint\limits_S \frac{P_n(\underline{x}_0)}{2\pi R} e^{i\omega R/c}\,\mathrm{d}S \qquad [1.77]$$

1.1.5.2. *Plane uniform transducer*

To calculate the diffraction impulse response of a plane transducer vibrating uniformly: $P_n(\underline{x}_0) = 1$, one approach is to note that in the integral [1.75], due to the term $\delta(t - R/c)$, only source points located at a distance $R = ct$ contribute to the field at point M and time t (Figure 1.11). These points are located on a circular arc, centered at the projection M_0 of the observation point M onto the plane of the disk. The radius ρ of this arc is such that:

$$\rho^2 + z^2 = R^2 \qquad \text{hence} \qquad \rho\,\mathrm{d}\rho = R\,\mathrm{d}R \qquad \text{and} \qquad \mathrm{d}\rho = \frac{\mathrm{d}R}{\sin\theta(R)} \qquad [1.78]$$

The surface element is $\mathrm{d}S = L(R)\,\mathrm{d}\rho$ where $L(R)$ is the length of the circular arc. Since $\delta(t - R/c) = c\delta(R - ct)$, the impulse response [1.75] takes the form Stepanishen (1971):

$$h(\underline{x}\,,t) = c\int_{R_1}^{R_2} \frac{\delta(R-ct)}{2\pi R}\frac{L(R)}{\sin\theta(R)}\,\mathrm{d}R \qquad [1.79]$$

where $R_1 = ct_1$ and $R_2 = ct_2$ denote, respectively, the minimum and maximum distances from the points on the disk to the observation point M. The integration amounts to replace R with ct:

$$\begin{cases} h(\underline{x}\,,t) = \dfrac{L(ct)}{2\pi t\sin[\theta(ct)]} & \text{for} \quad t_1 < t < t_2 \\ h(\underline{x}\,,t) = 0 & \text{otherwise} \end{cases} \qquad [1.80]$$

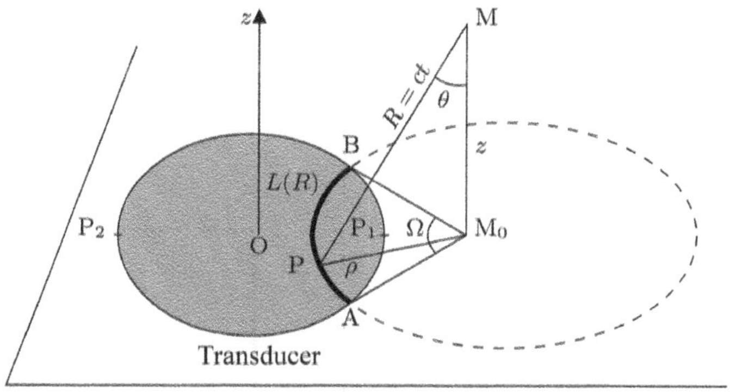

Figure 1.11. *At time t, the contribution to the acoustic field at point M comes from the source points on the circular arc AB, where the transducer plane intersects the sphere of radius $R = ct$ centered at M*

An even simpler expression is obtained by introducing the angle $\Omega(R = ct)$ which, from its center M_0 (Figure 1.11), subtends the circular arc AB of length:

$$L(ct) = \Omega(ct)\rho(ct) = \Omega(ct)ct\sin[\theta(ct)] \qquad [1.81]$$

The impulse response [1.80] takes the equivalent form:

$$\begin{cases} h(\underline{x}, t) = \dfrac{c}{2\pi}\Omega(ct) & \text{for} \quad t_1 < t < t_2 \\ h(\underline{x}, t) = 0 & \text{otherwise} \end{cases} \qquad [1.82]$$

In the case of the *uniform disk* of radius a, the field only depends on the variables r, z and t. For points located on the piston axis ($r = 0$) and for time t such that:

$$t_1 = z/c < t < t_2 = \sqrt{z^2 + a^2}/c \qquad [1.83]$$

the arcs that are equidistant from the observation point are complete circles, therefore $\Omega(ct) = 2\pi$:

$$h(z, r = 0, t) = c \quad \text{for} \quad t_1 < t < t_2 \qquad [1.84]$$

Since it is zero at other instants, the diffraction impulse response is a rectangular function whose duration:

$$\Theta(z) = t_2 - t_1 = \frac{1}{c}\left[\sqrt{z^2 + a^2} - z\right] \qquad [1.85]$$

decreases as the distance z increases (Figure 1.12).

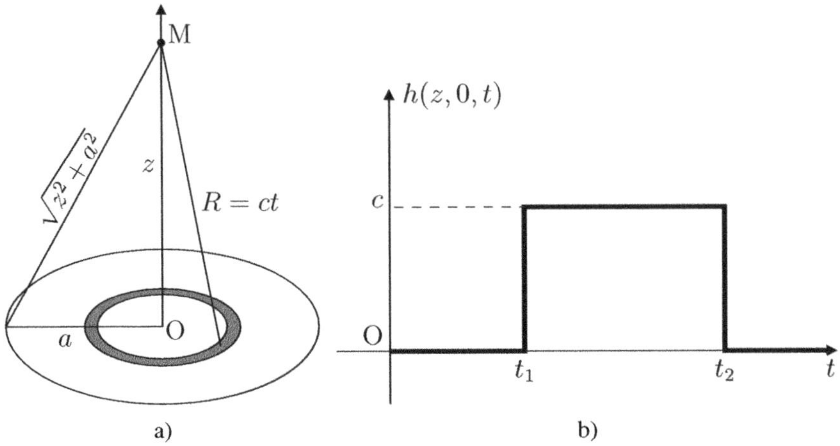

Figure 1.12. *Uniform disk of radius a. For a point M located on the z axis (a), the diffraction impulse response (b) is a rectangle of height c between times $t_1 = z/c$ and $t_2 = \sqrt{z^2 + a^2}/c$*

The *acoustic pressure* radiated by a circular disk, animated by a velocity $v_0(t)$, can be deduced from the time derivative of the velocity potential (equation [1.4]), that is, considering the convolution product:

$$p_a(\underline{x}, t) = \rho_0 \frac{\partial \phi(\underline{x}, t)}{\partial t} = \rho_0 v_0(t) \otimes \frac{\partial h(\underline{x}, t)}{\partial t} \qquad [1.86]$$

On the piston axis, the derivative of the impulse response (Figure 1.12(b)) consists of a positive delta at time t_1 and a negative delta at time t_2; we obtain:

$$p_a(z, 0, t) = \rho_0 c \left[v_0 \left(t - \frac{z}{c} \right) - v_0 \left(t - \frac{\sqrt{z^2 + a^2}}{c} \right) \right] \qquad [1.87]$$

Figure 1.13(a) shows that in the vicinity of the disk ($z << a$), the acoustic pressure is formed of a first term similar to the surface velocity (plane wave) and a second term, of opposite sign, whose arrival time corresponds to a path coming from the edge of the disk. This *edge wave* expresses the diffraction effects related to the finite radius a of the source. At a larger distance ($z = 2a$), the two waves begin to overlap (Figure 1.13(b)). The interference phenomena, which are responsible for the oscillations of the acoustic pressure observed on the piston axis in the sinusoidal case (Figure 1.4), disappear in the impulse regime, because the acoustic field received at one point and at a given time arises from a well-defined area on the surface of the transducer.

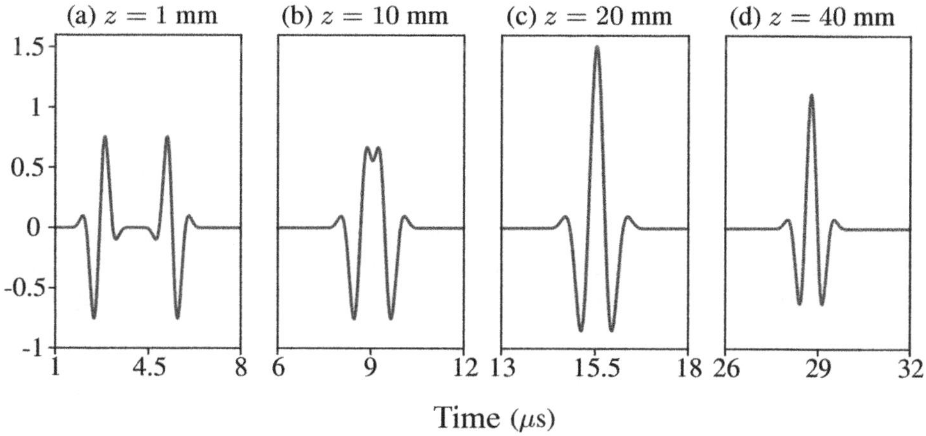

Figure 1.13. *Acoustic pressure emitted (in water) on the z axis of a plane transducer, of radius $a = 5$ mm and central frequency 1 MHz, excited by a short pulse. In the far field (d), the pressure is similar to the derivative of the velocity on the surface of the disk*

When the Fresnel approximation is valid, the duration Θ of the impulse response is inversely proportional to the distance z:

$$\sqrt{z^2 + a^2} \approx z + \frac{a^2}{2z} \qquad \text{hence} \qquad \Theta(z) \approx \frac{a^2}{2cz} \tag{1.88}$$

When $z^2 >> a^2$, the diffraction frequency response $H(\underline{x}\,, \omega)$, that is, the Fourier transform of $h(\underline{x}\,, t)$, undergoes a dilatation along the frequency axis, proportional to z.

In the *far field* (Fraunhofer zone), the diffraction impulse response tends toward a Dirac pulse, that is, considering the unit area of the Dirac delta function:

$$h(z, 0, t) \approx c\Theta(z)\delta\left(t - \frac{z}{c}\right) \approx \frac{a^2}{2z}\delta\left(t - \frac{z}{c}\right) \tag{1.89}$$

The velocity potential is similar to the velocity $v_0(t)$ of the points on the surface of the piston disk (formula [1.74]):

$$\phi(z, 0, t) = v_0(t) \otimes h(z, 0, t) \approx \frac{a^2}{2z} v_0\left(t - \frac{z}{c}\right) \tag{1.90}$$

and the acoustic pressure:

$$p_a(z,0,t) \approx \rho_0 \frac{a^2}{2z} \dot{v}_0 \left(t - \frac{z}{c}\right) \qquad [1.91]$$

is similar to the derivative of the surface velocity of the disk (Figure 1.13(d)).

Figure 1.14 shows that the acoustic pressure passes through a maximum at a distance such that the opposite contributions of plane wave and edge wave (equation [1.87]) are added, that is, when their path difference is equal to half a wavelength:

$$\sqrt{z^2 + a^2} - z = \frac{\lambda}{2} \qquad \text{hence} \qquad z = \frac{a^2}{\lambda} - \frac{\lambda}{4} \approx \frac{a^2}{\lambda} \qquad [1.92]$$

This "pseudo-focusing" distance corresponds to the position, z_1, of the last maximum (equation [1.41]), that is, the limit between the near field and the far field in the harmonic case.

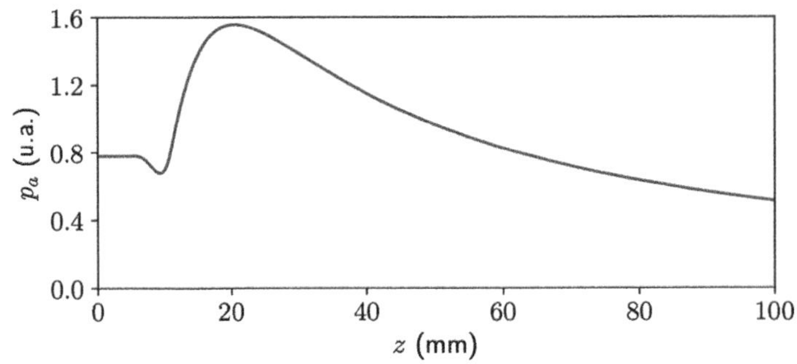

Figure 1.14. *Maximum of the acoustic pressure emitted in pulse regime by a plane transducer of radius $a = 5$ mm, and central frequency 1 MHz versus the z position on its axis*

REMARK.– The analysis of acoustic wave radiation by a transducer operating in the piston mode was conducted in two extreme cases, corresponding to the first two entries in Table 1.1:

– in the harmonic case, that is, for a sinusoidally vibrating piston with a constant amplitude;

– in the impulse regime, that is, for an infinitely short excitation (Dirac pulse).

We have shown that the radiated acoustic fields were very different (see, for example, Figures 1.4 and 1.14). In the ultrasonic domain, the need for a good axial resolution leads to the reduction of the time duration of the transmitted wave train. However, the center frequency and the bandwidth of the transducer impose a low limit on this duration (section 3.1.3). In practice, it varies from one period (very large bandwidth transducer) to a few tens of periods (narrow band transducer). The space-time evolution of the acoustic field, emitted in this transient regime, can then be calculated in two ways: either by summation of the contributions of the various spectral components (Fourier analysis) or, directly in the time domain, by convolution between the piston surface velocity and the diffraction impulse response.

1.1.5.3. *Concave uniform transducer*

In the far field, the acoustic beam emitted by a plane disk spreads due to diffraction: the acoustic intensity and lateral resolution decrease. A solution to improve these two parameters, essential to any imaging system, is to focus the acoustic beam either through a lens, like in optics, or by simply using a curved emitting surface.

The geometry of a focused transducer whose surface is a spherical cup is defined by its radius of curvature or focal length F, and by the radius a of its circular edge (Figure 1.15(a)). The wave radiated into the fluid by a concave source can be diffracted by its own surface, giving rise to a secondary radiation. Strictly speaking, the Rayleigh integral no longer applies. However, this secondary diffraction is negligible if the curvature of the surface is slight ($F >> a$). In this case, the angle θ_n, between the normal to the emitting surface and the z axis, is small ($\theta_n \leq \arctan(a/F)$). By equating the normal derivative ∂/∂_n with the axial derivative $\partial/\partial z$, expressions [1.75] and [1.77] for the Rayleigh integral are a good approximation of the radiated acoustic field.

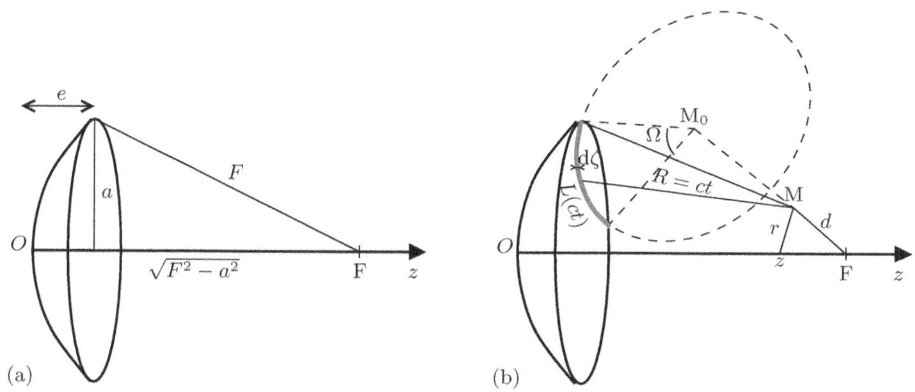

Figure 1.15. *Spherical cup, (a) focal length F, radius a, thickness e. (b) Calculation of the contribution to the radiated field at point M and time t*

1.1.5.3.1. Diffraction impulse response

In the case of a uniform vibrating surface: $P_n(\underline{x}_0) = 1$, the velocity potential $\phi(\underline{x}, t)$ and the diffraction impulse response $h(\underline{x}, t)$ are given by:

$$\phi(\underline{x}, t) = v_0(t) \otimes h(\underline{x}, t) \qquad \text{with} \qquad h(\underline{x}, t) = c \iint\limits_S \frac{\delta(R - ct)}{2\pi R} \, \mathrm{d}S \quad [1.93]$$

As for the plane disk, the surface element $\mathrm{d}S$ is equal to $L(R)\,\mathrm{d}\zeta$, where $L(R)$ is the length of the intersection of the spherical cup of radius F, with the sphere of radius $R = ct$ centered at the observation point M (Figure 1.15(b)), and $\mathrm{d}\zeta$ is the width of the band intercepted on the spherical cup. The result obtained by Penttinen and Luukkala (1976) is:

$$\begin{cases} h(z, r, t) = \dfrac{c}{2\pi} \dfrac{F}{d(z, r)} \Omega(ct) & \text{for} \quad t_1 < t < t_2 \\ h(z, r, t) = 0 & \text{otherwise} \end{cases} \quad [1.94]$$

where:

$$d(z, r) = \sqrt{r^2 + (F - z)^2} \quad [1.95]$$

is the distance between the observation point M(z, r) and the focus F(F, 0). t_1 and t_2 are the extreme arrival times of the disturbance emitted, at time $t = 0$, by all the points on the transducer. $\Omega(ct)$ is the angle that subtends the arc formed by the points on the cup located at a distance $R = ct$ from point M.

At any point on the axis ($r = 0$), since $\Omega(ct) = 2\pi$ and $d(z, 0) = |F - z|$, the diffraction impulse response takes the form of a rectangular function:

$$\begin{cases} h(z, 0, t) = c \dfrac{F}{|F - z|} & \text{for} \quad t_1 < t < t_2 \\ h(z, 0, t) = 0 & \text{otherwise} \end{cases} \quad [1.96]$$

Denoting the thickness of the spherical cup by:

$$e = F - \sqrt{F^2 - a^2} \quad [1.97]$$

the minimum and maximum transit times are

$$t_1 = \frac{z}{c} \quad \text{and} \quad t_2 = \frac{\sqrt{a^2 + (z - e)^2}}{c} \quad [1.98]$$

for disturbances emitted by the center and the edge of the cup, respectively. It must be noted that if $z < F$, then $t_1 < t_2$ and, conversely, if $z > F$, then $t_1 > t_2$. Considering expression [1.4], the acoustic pressure created by a surface velocity $v_0\delta(t)$ derives from the potential $\phi(t) = v_0 h(t)$:

$$p_a(z,0,t) = \rho_0 v_0 \frac{\partial h}{\partial t} = \rho_0 c v_0 \frac{F}{|F-z|}\left[\delta(t-t_1) - \delta(t-t_2)\right] \qquad [1.99]$$

that is, for its Fourier transform:

$$P_a(z,0,\omega) = Z_0 v_0 \frac{F}{|F-z|}\left[e^{i\omega t_1} - e^{i\omega t_2}\right] \qquad [1.100]$$

Given the value [1.98] of the limiting times t_1 and t_2 of the rectangle, the ratio of the acoustic intensity $I(z) = |P_a(z)|^2/2Z_0$ to the intensity $I_0 = Z_0 v_0^2/2$ on the transducer surface is given by:

$$\frac{I(z)}{I_0} = \frac{4F^2}{|F-z|^2}\sin^2\left[\frac{k}{2}\left(\sqrt{a^2+(z-e)^2} - z\right)\right] \qquad [1.101]$$

Figure 1.16 represents the variations of the acoustic intensity $I(z)/I_0$ normalized by the intensity I_0 on the surface of a transducer of focal length $F = 50$ mm, radius $a = 10$ mm, that is, $e = 1$ mm, excited at a frequency $f = 2.4$ MHz, so that $ke = 10$. Before the focus, the oscillations resulting from the interference between waves emitted by the edge and the center of the cup are much smaller than in the case of a plane disk. The maximum of the relative intensity is located a little before the focus; beyond this point the acoustic intensity decreases rapidly with the distance.

When the observation point goes toward the focus ($z \to F$), the duration of the response tends to zero, while its amplitude tends to infinity. The diffraction impulse response converges to a Dirac function, whose intensity A is equal to the limit, when z approaches F, of the area of the rectangle of height $cF/|F-z|$ and duration $|t_1 - t_2|$:

$$A = \lim_{z\to F}\left[F\frac{|ct_1 - ct_2|}{|F-z|}\right] \qquad [1.102]$$

As to the first order in $(z-F)$:

$$\sqrt{a^2+(z-e)^2} \approx F + \frac{z-F}{F}\sqrt{F^2-a^2} \quad \text{and} \quad ct_1 - ct_2 \approx \frac{z-F}{F}e \qquad [1.103]$$

the area A of the rectangle is equal to the thickness e of the spherical cup, so that:

$$h(F,0,t) = e\delta\left(t - \frac{F}{c}\right) \qquad [1.104]$$

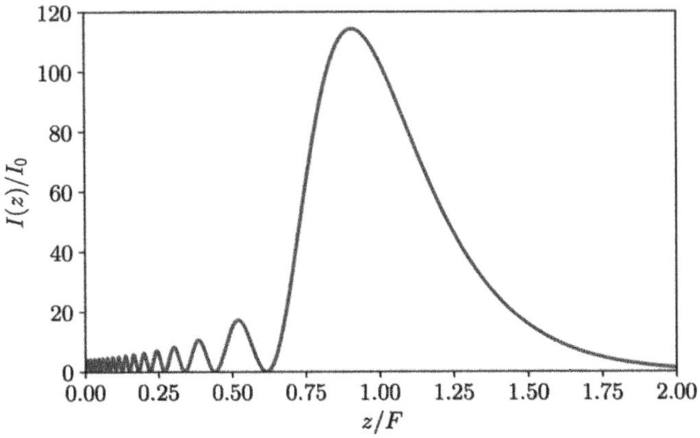

Figure 1.16. *Normalized intensity along the axis of a transducer of focal length $F = 50$ mm and radius $a = 10$ mm*

At the transducer focus, the velocity potential is a replica of the velocity $v_0(t)$ of the points on the emitting surface: $\phi(F, 0, t) = ev_0(t - F/c)$. The velocity along z is proportional to the derivative of $v_0(t)$ delayed by F/c:

$$v_z(F, 0, t) = -\left.\frac{\partial \phi}{\partial z}\right|_{z=F} = \frac{e}{c}\dot{v}_0\left(t - \frac{F}{c}\right) \qquad [1.105]$$

Beyond the focus, the shape of the velocity no longer changes and only its amplitude decreases due to the divergence of the beam. The particle velocity (or acoustic pressure) undergoes a time-derivative as it passes through the focus of the transducer.

As mentioned at the beginning of this section, the advantage of using a focused transducer is to achieve better performances in acoustic detection or ultrasonic imaging. In the following, we evaluate these improvements in terms of sensitivity and spatial resolution.

1.1.5.3.2. Acoustic intensity at the focus, diameter of the focal spot

The acoustic pressure is given by relation [1.4] and the diffraction frequency response $H(F, 0, \omega)$ is the Fourier transform of the impulse response [1.104]:

$$P_a(F, 0, \omega) = -i\omega\rho_0\Phi(F, 0, \omega) \qquad \text{and} \qquad H(F, 0, \omega) = e\, e^{i\omega F/c} \qquad [1.106]$$

Since, in the frequency domain $\Phi(F,0,\omega) = H(F,0,\omega)V_0(\omega)$, the modulus of the acoustic pressure at the transducer focus is equal to:

$$|P_a(F,0,\omega)| = \omega\rho_0 e|V_0(\omega)| \qquad [1.107]$$

On the surface of the spherical cup, $p_a(0,0,t) = \rho_0 c v_0(t)$, hence $P_a(0,0,\omega) = \rho_0 c V_0(\omega)$, and the ratio of the acoustic intensities is equal to the ratio of the square of the acoustic pressure moduli:

$$\frac{I_F}{I_0} = \left|\frac{P_a(F,0,\omega)}{P_a(0,0,\omega)}\right|^2 = (ke)^2 \qquad [1.108]$$

Assuming that $a^2 << F^2$, the thickness of the spherical cup can be written as $e \approx a^2/2F$ and the gain in intensity:

$$\frac{I_F}{I_0} = \left(\frac{\pi a^2}{\lambda F}\right)^2 \qquad [1.109]$$

is larger than unity if the dimensionless parameter $\lambda F/a^2$ is less than π, that is, if the focal length F is smaller than the Rayleigh distance (Kino 1987), defined by:

$$z_R = \frac{\pi a^2}{\lambda} \qquad \text{with} \qquad a >> \lambda \qquad [1.110]$$

Energy concentration is only possible in the near field, beyond this limit the natural divergence of the acoustic beam prevails over the focusing effect due to the curvature of the spherical transducer.

In the piston mode, the mean power emitted by the entire surface of the transducer is $\pi a^2 I_0$. By defining the diameter of the acoustic beam at the focus by the value $2w$, such that half the emitted power passes through the interior of a circle of radius w, we obtain:

$$\pi w^2 I_F = \frac{1}{2}\pi a^2 I_0 \qquad \text{hence} \qquad 2w = a\sqrt{2\frac{I_0}{I_F}} = \frac{\sqrt{2}}{\pi}\frac{\lambda F}{a} \qquad [1.111]$$

The width of the acoustic beam at –3 dB is thus written as:

$$2w = \frac{2\sqrt{2}}{\pi}\frac{\lambda F}{D} \approx \frac{\lambda F}{D} \qquad [1.112]$$

where D and F/D are, respectively, the diameter and the F-number of the transducer.

1.2. Generation of elastic waves by a surface source

In order to examine the radiation of bulk and surface elastic waves by sources distributed on the surface of a solid, we use Fourier analysis and the formalism of the mixed matrix, discussed in Appendix 2. The approach is as follows:

– expression of stresses in the harmonic case as a function of the amplitudes of the rising and falling waves;

– matrix inversion to calculate, using the emission matrix, the amplitudes of the emitted waves as functions of the forces applied on the solid surface;

– calculation of the inverse Fourier transform to obtain the spatiotemporal response. In the near field, the inversion is carried out numerically; in the far field, approximate analytical expressions for the mechanical displacements at the arrival time of the bulk wave fronts and the Rayleigh wave fronts are obtained using the stationary phase method.

The method is applied to the case of an infinitely long and thin line source located along the x_3 axis, on the surface of a solid occupying the half-space $x_2 > 0$; the impulsive forces are either normal or parallel to the surface (Figure 1.17).

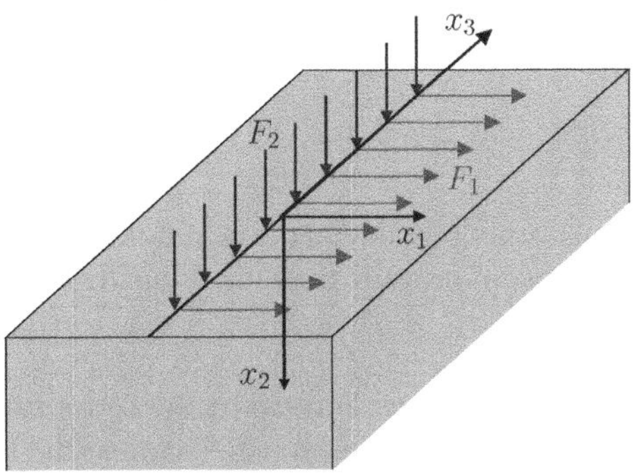

Figure 1.17. *Arrangement of the line source on the surface $x_2 = 0$ of the solid. The normal force density per unit length is F_2. The tangential force density is F_1. For a color version of this figure, see www.iste.co.uk/royer/waves2.zip*

The mixed matrix $\boldsymbol{M}$ relates the amplitude $\underline{a}$ of the upward bulk waves and the amplitude $\underline{b}$ of the downward waves to the mechanical traction $T_k = \sigma_{k2}$ and the displacements u_i on the solid surface. The physical meaning of the sub-matrices is given in Appendix 2. Here, in the absence of any rising wave ($\underline{a} = \underline{0}$) and according to definition [A2.1], only the matrices $\boldsymbol{E}$ and $\boldsymbol{Y}$ come into play to express, in terms of the mechanical traction T_k, the amplitudes b_J of the emitted bulk waves and the components u_i of the displacement on the free surface:

$$\underline{b} = \boldsymbol{E}\underline{T} \qquad \text{and} \qquad \underline{u} = \boldsymbol{Y}\underline{T} \tag{1.113}$$

To arrive at a two-dimensional problem, the material, *a priori* anisotropic, is assumed to have an orthotropic symmetry with a binary axis parallel to x_3. The propagation of QL and QT waves, polarized in the x_1x_2 plane, is then decoupled from the propagation of the TH wave, polarized along x_3 (Volume 1, section 2.2.2.1). The slowness component s_3 is equal to zero and the propagation direction in the sagittal plane is defined by the ratio $m = s_2/s_1$ of the two other components. By taking s_1 as variable, the Christoffel equation is a fourth-degree polynomial equation in s_2, whose coefficients are real. This admits four real or complex roots. Each real root corresponds to a propagative bulk wave transporting energy. The direction of the power flux is given by the normal to the slowness surface. The complex roots appear in conjugated pairs and only the two solutions decreasing with x_2 must be retained as components of the Rayleigh wave. These two acceptable roots (s_{2J} with $J = 1,2$) and all the elements of the mixed matrix are expressed as a function of a single variable: the slowness component s_1.

We first examine the generation of bulk waves polarized in the sagittal plane and then the generation of Rayleigh waves. In the harmonic case, an integral expression gives the mechanical displacement created in an elastic solid by a distribution of mechanical traction $T_k(x_1)\exp(-i\omega t)$ applied to the free surface of the solid at $x_2 = 0$. The relation [1.113] shows that the components E_{Jk} of the emission matrix are the coefficients of generation of the bulk elastic wave J by a tangential force ($k = 1$) or a normal force ($k = 2$). A surface excitation, concentrated along the x_3 axis: $T_k(x_1) = F_k\delta(x_1)$ can be developed as the sum of spatial harmonics:

$$T_k(x_1) = \frac{F_k}{2\pi}\int_{-\infty}^{\infty} e^{ik_1x_1}\,\mathrm{d}k_1 \tag{1.114}$$

As the mechanical traction $T_k = \sigma_{k2}$ is expressed in N/m^2, F_k is expressed in N/m: this is the force density per unit length along x_3. In the plane of abscissa x_2, the complex amplitude b_J of the bulk wave J created by the uniform elementary force $F_k\,\mathrm{d}k_1$, parallel to the x_k axis is:

$$\mathrm{d}b_J = E_{Jk}F_k e^{ik_{2J}x_2}\,\mathrm{d}k_1 \tag{1.115}$$

By introducing, for the mode J, the components $s_1 = k_1/\omega$ and $s_{2J} = k_{2J}/\omega$ of the slowness vector, the total amplitude emitted by the line source is given by the integral:

$$b_J(x_1, x_2, \omega) = \frac{\omega}{2\pi} F_k \int_{-\infty}^{\infty} E_{Jk}(s_1) e^{i\omega(s_1 x_1 + s_{2J} x_2)} \, \mathrm{d}s_1 \qquad [1.116]$$

1.2.1. *Solid with orthotropic symmetry*

We begin by developing the method in the general case of a medium with orthotropic symmetry, before applying it to an isotropic solid, and then to a material of cubic symmetry. The propagation equations for the QL and QT plane waves polarized in the x_1x_2 plane of an orthotropic solid were developed in sections 2.5.5 and 3.1.2.2 of Volume 1. With a mechanical displacement:

$$u_l = b q_l e^{i(k_1 x_1 + k_2 x_2 - \omega t)}, \qquad l = 1, 2 \qquad [1.117]$$

they are written in matrix form (Volume 1, system [3.43] with $m = k_2/k_1$):

$$\begin{pmatrix} C_{11}k_1^2 + C_{66}k_2^2 - \rho\omega^2 & (C_{12} + C_{66})k_1k_2 \\ (C_{12} + C_{66})k_1k_2 & C_{66}k_1^2 + C_{22}k_2^2 - \rho\omega^2 \end{pmatrix} \begin{pmatrix} q_1 \\ q_2 \end{pmatrix} = \begin{pmatrix} 0 \\ 0 \end{pmatrix} \qquad [1.118]$$

Dividing by ω^2 reveals the slownesses $s_1 = k_1/\omega$ and $s_2 = k_2/\omega$. The characteristic equation:

$$\left(C_{11}s_1^2 + C_{66}s_2^2 - \rho\right)\left(C_{66}s_1^2 + C_{22}s_2^2 - \rho\right) - \left(C_{12} + C_{66}\right)^2 s_1^2 s_2^2 = 0 \qquad [1.119]$$

is a biquadratic equation in s_2^2:

$$\alpha_4 s_2^4 - \alpha_2 s_2^2 + \alpha_0 = 0 \qquad [1.120]$$

whose coefficients α_0, α_2 and α_4 are given by relations [2.172] of Volume 1. The two acceptable roots ($J = 1, 2$) corresponding to the quasi-longitudinal wave (s_{21}) and quasi-transverse wave (s_{22}) depend on the slowness s_1 through the intermediary of coefficients α_2 and α_0:

$$s_{21}^2 = \frac{\alpha_2 - \sqrt{\alpha_2^2 - 4\alpha_0\alpha_4}}{2\alpha_4} \quad \text{and} \quad s_{22}^2 = \frac{\alpha_2 + \sqrt{\alpha_2^2 - 4\alpha_0\alpha_4}}{2\alpha_4} \qquad [1.121]$$

The polarization vector components for each wave, q_{1J} and q_{2J}, are related by:

$$\left(C_{11}s_1^2 + C_{66}s_{2J}^2 - \rho\right) q_{1J} + (C_{12} + C_{66}) s_1 s_{2J} q_{2J} = 0 \qquad [1.122]$$

where s_{2J} ($J = 1, 2$) are the two positive roots of equations [1.121]. The stresses $T_i = \sigma_{i2}$ on the surface $x_2 = 0$ are expressed using the coefficients B_{iJ} (Appendix 2) and the amplitudes b_J of QL and QT waves:

$$T_i = ik_1 \sum_{J=1}^{2} B_{iJ} b_J = i\omega s_1 \sum_{J=1}^{2} B_{iJ} b_J \qquad i = 1, 2 \qquad [1.123]$$

The expressions for these coefficients, identical to those for coefficients A_{iJ}, are given as functions of the parameter $m_J = s_{2J}/s_1$ by relations [2.150] in Volume 1, that is:

$$\begin{cases} s_1 B_{1J} = C_{66}\left(s_{2J} q_{1J} + s_1 q_{2J}\right) \\ s_1 B_{2J} = C_{12} s_1 q_{1J} + C_{22} s_{2J} q_{2J} \end{cases} \qquad [1.124]$$

The inversion of the linear system [1.123], that is, of matrix $\boldsymbol{B}$, provides the four components E_{Jk} of the emission matrix $\boldsymbol{E}$, required for the calculation of the integral [1.116], giving the amplitude b_J of wave J created by the force density F_k:

$$\underline{b} = \boldsymbol{E}\underline{T} \quad \text{with} \quad \boldsymbol{E} = \frac{1}{i\omega s_1 \left(B_{11}B_{22} - B_{12}B_{21}\right)} \begin{pmatrix} B_{22} & -B_{12} \\ -B_{21} & B_{11} \end{pmatrix} \qquad [1.125]$$

Since the amplitudes of the partial waves, that is, the components of vector $\underline{b}$, are obtained in the Fourier domain, the components of the displacement field in the real space are calculated by numerical inverse Fourier transforms. However, the presence of singularities associated with the poles or the branch points makes the inversion difficult. The introduction of a small imaginary part in the angular frequency allows us to move away from these singularities while keeping their significant influence (Weaver *et al.* 1996).

The generation of elastic waves by a line source in a duralumin half-space is illustrated in Figure 1.18. The source, located at $(x_1, x_2) = (0, 0)$, is a very short impulse. Four types of waves are visible: the first two are longitudinal (L) and transverse (T) bulk waves, which propagate with a cylindrical wave front, diverging from the line source. Two Rayleigh (R) waves propagate on the free surface on both sides of the source. Finally, two head waves (H) can also be observed on either side of the source. These head waves are transverse waves resulting from the conversion of the grazing longitudinal wave beyond the critical angle $\theta_c = \arcsin(V_T/V_L)$ (see Chapter 2 of Volume 1). Due to this conversion, the grazing longitudinal wave radiates energy toward the inside of the material, implying that it is attenuated during its propagation.

The generation of elastic waves by an impulse line source in a copper half-space is illustrated in Figure 1.19. The comparison of this map at time $t = 40$ μs with the wave surfaces in Figure 1.13(b) of Volume 1 clearly shows quasi-longitudinal and quasi-transverse wave fronts.

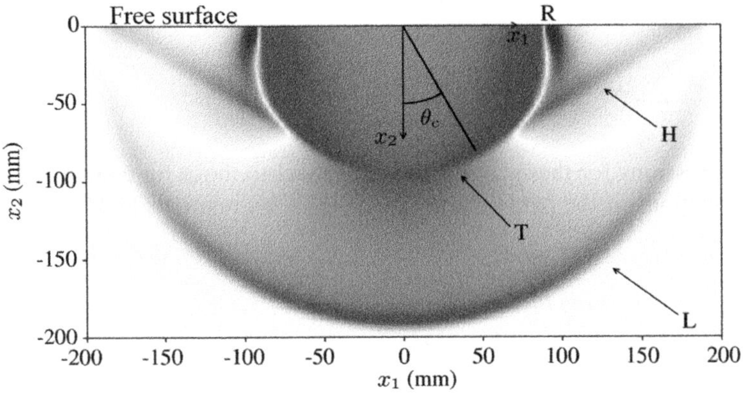

Figure 1.18. *Isotropic solid. Map of the amplitude of the displacement generated by an impulse line source and wave fronts. For a color version of this figure, see www.iste.co.uk/royer/waves2.zip*

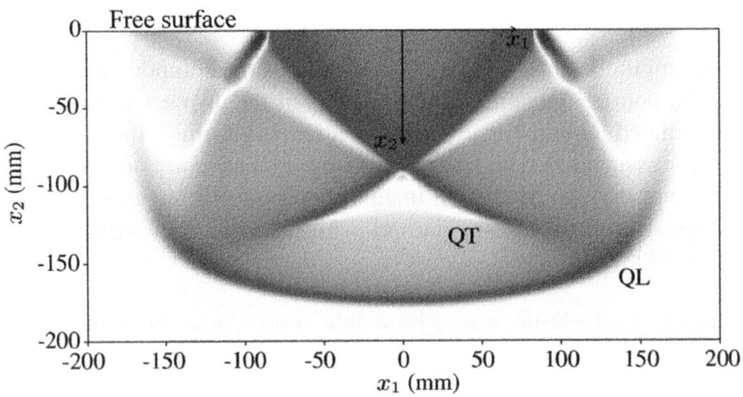

Figure 1.19. *Anisotropic solid. Map at time $t = 40$ μs after the application of an impulsive force, showing the amplitude of the displacement generated by a line source in a copper crystal with cubic symmetry. For a color version of this figure, see www.iste.co.uk/royer/waves2.zip*

1.2.2. *Far field*

In this section, the integral [1.116] is evaluated by the stationary phase method. This approximation is valid in the far field and in the vicinity of the wave fronts of bulk waves, when they propagate independently. The directivity functions are plotted for an isotropic solid. In the case of an anisotropic material, the focusing effect due to the variations in curvature of the slowness surface is highlighted. Then, we calculate the contribution of the Rayleigh wave to the mechanical displacement of the solid surface.

1.2.2.1. *Bulk waves: directivity diagrams*

When $k_1 x_1$ is very large compared to unity, the phase variations:

$$\varphi(s_1) = \omega(s_1 x_1 + s_2 x_2) \qquad \text{with} \qquad s_2 = s_{2J}(s_1) \qquad [1.126]$$

cause very rapid oscillations (as functions of s_1) of the exponential $e^{i\varphi(s_1)}$, which contribute little to the integral [1.116]. The major part of its value comes from the interval around the slowness s_1^0 where the phase $\varphi(s_1)$ is stationary, that is, for which the derivative $\mathrm{d}\varphi/\mathrm{d}s_1$ cancels. The second-order development is a good approximation of the phase in the vicinity of s_1^0:

$$\varphi(s_1) = \varphi_0 + \frac{1}{2}\varphi_0''\left(s_1 - s_1^0\right)^2 \text{ where } \varphi_0 = \varphi\left(s_1^0\right) \text{ and } \varphi_0'' = \left.\frac{\partial^2\varphi}{\partial s_1^2}\right|_{s_1^0} \qquad [1.127]$$

Let us carry over this development into [1.116] and by assuming that the variations in factor $E_{Jk}(s_1)$ are slow compared to the variations in the exponential, we get:

$$\begin{aligned} I &= \int_{-\infty}^{\infty} E_{Jk}(s_1) e^{i\varphi(s_1)}\,\mathrm{d}s_1 \\ &= e^{i\varphi_0} E_{Jk}(s_1^0) \int_{-\infty}^{\infty} \exp\left[i\frac{1}{2}\varphi_0''\left(s_1 - s_1^0\right)^2\right] \mathrm{d}s_1 \end{aligned} \qquad [1.128]$$

By changing the variable:

$$|\varphi_0''|\left(s_1 - s_1^0\right)^2 = 2\pi x^2 \qquad \longrightarrow \qquad \mathrm{d}s_1 = \sqrt{\frac{2\pi}{|\varphi_0''|}}\,\mathrm{d}x \qquad [1.129]$$

and since $\varphi_0'' = \varepsilon|\varphi_0''|$, we obtain:

$$I = e^{i\varphi_0} E_{Jk}\left(s_1^0\right)\sqrt{\frac{2\pi}{|\varphi_0''|}}\int_{-\infty}^{\infty} e^{i\pi\varepsilon x^2}\,\mathrm{d}x \qquad \text{with} \qquad \varepsilon = \mathrm{sign}\left(\varphi_0''\right) \qquad [1.130]$$

As:

$$\int_{-\infty}^{\infty} e^{i\pi\varepsilon x^2}\,\mathrm{d}x = e^{i\varepsilon\frac{\pi}{4}} \qquad [1.131]$$

the amplitude of the wave J emitted by the source is given by equation [1.116]:

$$b_J(x_1, x_2, \omega) = \omega F_k \frac{E_{Jk}\left(s_1^0\right)}{\sqrt{2\pi|\varphi_0''|}} e^{i(\varphi_0 + \varepsilon\pi/4)} \qquad [1.132]$$

1.2.2.1.1. Isotropic solid

As $s_1 = s_J \sin\theta$ and $s_2 = s_J \cos\theta$, at the fixed observation point M(x_1, x_2), the derivative of the phase:

$$\left.\frac{\partial\varphi}{\partial s_1}\right|_{x_i} = \omega\left(x_1 + x_2\frac{\mathrm{d}s_2}{\mathrm{d}s_1}\right) = \omega(x_1 - x_2\tan\theta) \qquad [1.133]$$

vanishes for the angle θ_0, so that $\tan\theta_0 = x_1/x_2$ and the second derivative is negative:

$$\varphi'' = \left(\frac{\mathrm{d}^2\varphi}{\mathrm{d}s_1^2}\right)_{x_i} = -\frac{\omega x_2}{\cos^2\theta}\frac{\mathrm{d}\theta}{\mathrm{d}s_1} = -\frac{\omega r}{s_J\cos^2\theta} \qquad [1.134]$$

By introducing the wave number $k_J = \omega s_J$, we get:

$$\varphi_0 = k_J r \qquad \text{and} \qquad \varphi_0'' = -\frac{k_J r}{s_J^2\cos^2\theta} \qquad [1.135]$$

With $\varepsilon = -1$, equation [1.132] gives the amplitude of the longitudinal ($J = 1$) and transverse ($J = 2$) displacements generated by a linear distribution of tangential ($k = 1$) or normal ($k = 2$) forces:

$$b_J(r, \theta, \omega) = 2i\omega F_k s_J \cos\theta E_{Jk}(s_J \sin\theta)\frac{e^{ik_J r - i\pi/4}}{2i\sqrt{2\pi k_J r}} \qquad [1.136]$$

This amplitude decreases in $1/\sqrt{r}$, as expected for cylindrical waves emitted by a line source. The last term of the right member is, in the two-dimensional case, the far field approximation of the spectral Green's function of the Helmholtz equation:

$$G(r, \omega) = \frac{1}{4i}H_0^{(1)}(kr)e^{-i\omega t} \qquad [1.137]$$

Taking into account its asymptotic behavior, the zero-order Hankel function of first kind $H_0^{(1)}$ is the appropriate solution for a divergent wave and a temporal factor in $e^{-i\omega t}$:

$$H_0^{(1)}(kr) \approx \sqrt{\frac{2}{\pi kr}}e^{ikr - i\pi/4} \qquad \text{for} \qquad kr >> 1 \qquad [1.138]$$

Given the expression for the time Green's function, that is, the Fourier transform of $G(r,\omega)$:

$$g(r,t) = \frac{H(t-sr)}{2\pi\sqrt{t^2-s^2r^2}} \qquad [1.139]$$

where $H(t)$ is the unit step (Heaviside) function, the instantaneous displacement is given by:

$$u_J(r,\theta,t) = i\omega\frac{F_k}{\pi}s_J\cos\theta E_{Jk}(s_J\sin\theta)\frac{H(t-s_Jr)}{\sqrt{t^2-s_J^2r^2}} \qquad [1.140]$$

Normal force (k = 2)

The coefficients $E_{12}(s_1)$ and $E_{22}(s_1)$ are calculated in Appendix 2:

$$E_{12}(s_1) = \frac{(s_T^2-2s_1^2)s_L}{i\omega\mu\Delta(s_1)} \quad \text{and} \quad E_{22}(s_1) = \frac{2s_1s_{2L}s_T}{i\omega\mu\Delta(s_1)} \qquad [1.141]$$

where $\Delta(s_1)$ is the Rayleigh determinant:

$$\Delta(s_1) = \left(2s_1^2-s_T^2\right)^2 + 4s_1^2s_{2L}s_{2T} \qquad [1.142]$$

– In the case of a *longitudinal wave*: $s_1 = s_L\sin\theta$ and $s_{2L} = s_L\cos\theta$, the Rayleigh determinant is expressed by:

$$\begin{aligned}\Delta(s_L\sin\theta) &= s_L^4\left[(\kappa^{-2}-2\sin^2\theta)^2 + 2\sin\theta\sin 2\theta\sqrt{\kappa^{-2}-\sin^2\theta}\right] \\ &= s_L^4\Delta_L(\theta)\end{aligned} \qquad [1.143]$$

with $\kappa = s_L/s_T$. By replacing s_1 by $s_L\sin\theta$ in the numerator of E_{12}, we get:

$$i\omega E_{12}(s_L\sin\theta) = \frac{\kappa^{-2}-2\sin^2\theta}{\mu s_L\Delta_L(\theta)} \qquad [1.144]$$

and the displacement in the far field [1.140] is equal to:

$$u_L(r,\theta,t) = \frac{F_2}{\pi\mu}D_{2L}(\theta)\frac{H(t-s_Lr)}{\sqrt{t^2-s_L^2r^2}}. \qquad [1.145]$$

$D_{2L}(\theta)$ is the dimensionless directivity factor for a normal force:

$$D_{2L}(\theta) = \frac{\cos\theta(\kappa^{-2}-2\sin^2\theta)}{(\kappa^{-2}-2\sin^2\theta)^2 + 2\sin\theta\sin 2\theta\sqrt{\kappa^{-2}-\sin^2\theta}} \qquad [1.146]$$

– In the case of a *transverse wave* $s_1 = s_T \sin\theta$ and $s_{2T} = s_T \cos\theta$, the Rayleigh determinant is expressed by:

$$\begin{aligned}\Delta(s_T \sin\theta) &= s_T^4 \left[(1 - 2\sin^2\theta)^2 + 2\sin\theta \sin 2\theta \sqrt{\kappa^2 - \sin^2\theta}\right] \\ &= s_T^4 \Delta_T(\theta)\end{aligned} \qquad [1.147]$$

By replacing s_1 by $s_T \sin\theta$ in the numerator of E_{22}, we obtain:

$$i\omega E_{22}(s_T \sin\theta) = \frac{2\sin\theta\sqrt{\kappa^2 - \sin^2\theta}}{\mu s_T \Delta_T(\theta)} \qquad [1.148]$$

The transverse displacement in the far field is given by expression [1.145], by replacing $D_{2L}(\theta)$ with:

$$D_{2T}(\theta) = \frac{\sin 2\theta\sqrt{\kappa^2 - \sin^2\theta}}{\cos^2 2\theta + 2\sin\theta \sin 2\theta\sqrt{\kappa^2 - \sin^2\theta}} \qquad [1.149]$$

The directivity patterns $D_{2L}(\theta)$ and $D_{2T}(\theta)$ are plotted in Figure 1.20 in the case of duralumin ($\kappa = V_T/V_L = 0.5$) and for a material of Poisson's ratio $\nu = 0.2$. The radiation of the longitudinal wave is maximum at the center. The source is omnidirectional but not isotropic. For transverse waves, the source is more directional. Maximum energy is radiated at an angle close to 35°.

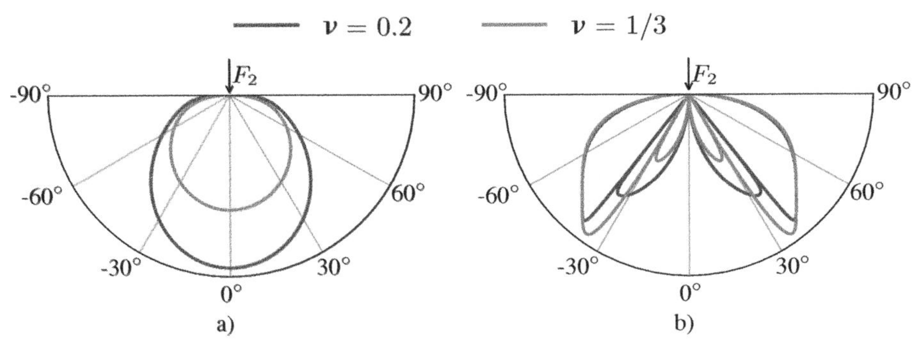

Figure 1.20. *Normal force. Radiation patterns for a line source: a) longitudinal wave; b) transverse wave for two values of the Poisson's ratio: $\nu = 0.2$ and $\nu = 1/3$. For a color version of this figure, see www.iste.co.uk/royer/waves2.zip*

ORDER OF MAGNITUDE.– The amplitude of the mechanical displacement, proportional to the ratio of the force density F_2 (N/m) to the shear modulus μ (N/m^2), is expressed in meters. In the ultrasound domain, the displacements are of the order of a few nanometers. For example, with constants close to those of duralumin: $\mu = 25$ GPa, $\kappa = 0.5$ and $F_2 = 1$ N/cm, we obtain $D_{2L}(0) = \kappa^2 = 0.25$ and $u_L = 1$ nm.

Tangential force (k = 1)

The elements $E_{11}(s_1)$ and $E_{21}(s_1)$ of the emission matrix [A2.24] are:

$$E_{11}(s_1) = \frac{2s_1 s_{2T} s_L}{i\omega\mu\Delta(s_1)} \quad \text{and} \quad E_{21}(s_1) = \frac{(2s_1^2 - s_T^2)s_T}{i\omega\mu\Delta(s_1)} \tag{1.150}$$

and the Rayleigh determinant is still expressed by:

$$\Delta(s_L \sin\theta) = s_L^4 \Delta_L(\theta) \quad \text{and} \quad \Delta(s_T \sin\theta) = s_T^4 \Delta_T(\theta) \tag{1.151}$$

Replacing s_1 by $s_L \sin\theta$ in the numerator of E_{11}, the *longitudinal displacement* created in the far field by the tangential force distribution of density F_1 is given by:

$$u_L(r,\theta,t) = \frac{F_1}{\pi\mu} D_{1L}(\theta) \frac{H(t - s_L r)}{\sqrt{t^2 - s_L^2 r^2}} \tag{1.152}$$

where the directivity factor $D_{1L}(\theta)$ has the same denominator as $D_{2L}(\theta)$:

$$D_{1L}(\theta) = \frac{\sin 2\theta\sqrt{\kappa^{-2} - \sin^2\theta}}{\left(\kappa^{-2} - 2\sin^2\theta\right)^2 + 2\sin\theta\sin 2\theta\sqrt{\kappa^{-2} - \sin^2\theta}} \tag{1.153}$$

Similarly, by replacing s_1 by $s_T \sin\theta$ in E_{21}, the *transverse displacement* is given by expression [1.152] by changing $D_{1L}(\theta)$ in:

$$D_{1T}(\theta) = \frac{\cos\theta\cos 2\theta}{\cos^2 2\theta + 2\sin\theta\sin 2\theta\sqrt{\kappa^2 - \sin^2\theta}} \tag{1.154}$$

The denominator of $D_{1T}(\theta)$ is identical to that of $D_{2T}(\theta)$.

Figure 1.21(a) shows that there is no radiation in the form of a longitudinal wave along the normal to the surface. The emission is maximum in a direction depending on the ratio $\kappa = V_T/V_L$, which is close to 65° in the case of duralumin ($\nu \approx 1/3$). For transverse waves (Figure 1.21(b)), maximum energy is radiated at an angle close to 30°. Emission is always zero for θ = 45° due to the cancellation of the term $\cos 2\theta$ in formula [1.154]. The main and secondary lobes have opposite polarities.

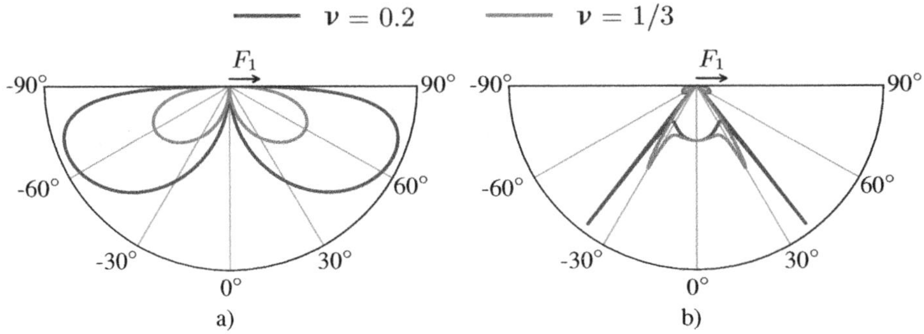

Figure 1.21. *Tangential force. Radiation patterns for a line source: a) longitudinal wave; b) transverse wave for two values of Poisson's ratio: $\nu = 0.2$ and $\nu = 1/3$. For a color version of this figure, see www.iste.co.uk/royer/waves2.zip*

Since the ratio $\kappa^{-2} = (V_L/V_T)^2$ is greater than 2 for any isotropic solid, the directivity functions D_{1L} and D_{2L} are real for any θ. On the other hand, the functions D_{1T} and D_{2T} are real for θ smaller than the critical angle $\theta_c = \arcsin(V_T/V_L)$ and complex for $\theta > \theta_c$.

1.2.2.1.2. Anisotropic solid

The energy velocity vector $\underline{V}^e$ is at all points normal to the slowness surface: $\underline{V}^e.\,\mathrm{d}\underline{s} = 0$. Assuming that the sagittal plane is a symmetry plane, $\underline{V}^e$ is contained in this plane and the angle θ^e of the acoustic ray, with respect to the normal x_2, is given by $\tan\theta^e = -\,\mathrm{d}s_2/\,\mathrm{d}s_1$. For a given source–receiver configuration, the derivative of the phase:

$$\left.\frac{\partial\varphi}{\partial s_1}\right|_{x_i} = \omega\left(x_1 + x_2\frac{\mathrm{d}s_2}{\mathrm{d}s_1}\right) = \omega(x_1 - x_2\tan\theta^e) \qquad [1.155]$$

vanishes if $\tan\theta^e = x_1/x_2$. Consequently, the value s_1^0 for which the phase is stationary is such that the acoustic ray is parallel to the direction joining the emission point O to the observation point M (Figure 1.22).

The associated point L on the slowness curve of mode J defines the propagation angle θ through:

$$s_1^0 = s_J(\theta)\sin\theta \qquad \text{and} \qquad s_2^0 = s_J(\theta)\cos\theta \qquad [1.156]$$

Setting $OM = r$: $x_1 = r\sin\theta^e$, $x_2 = r\cos\theta^e$, the value of the phase $\varphi_0 = \varphi(s_1^0)$ is:

$$\varphi_0 = \omega s_J(\theta) r(\sin\theta\sin\theta^e + \cos\theta\cos\theta^e) = k_J r\cos\psi \qquad [1.157]$$

where $\psi = \theta^e - \theta$ is the deviation angle between the energy velocity vector and the propagation direction. Let us calculate the second derivative from the expression [1.155] of the first derivative:

$$\varphi'' = \left(\frac{\mathrm{d}^2\varphi}{\mathrm{d}s_1^2}\right)_{x_i} = -\frac{\omega r}{\cos\theta^e}\frac{\mathrm{d}\theta^e}{\mathrm{d}s_1} \qquad [1.158]$$

By using the relation $\mathrm{d}s/\mathrm{d}\theta = -s\tan\psi$ (Figure 1.22) to evaluate the derivative of $s_1 = s(\theta)\sin\theta$, we obtain:

$$\frac{\mathrm{d}s_1}{\mathrm{d}\theta^e} = \left(\frac{\mathrm{d}s}{\mathrm{d}\theta}\sin\theta + s(\theta)\cos\theta\right)\frac{\mathrm{d}\theta}{\mathrm{d}\theta^e} = s(\theta)\frac{\cos\theta^e}{\cos\psi}\frac{\mathrm{d}\theta}{\mathrm{d}\theta^e} \qquad [1.159]$$

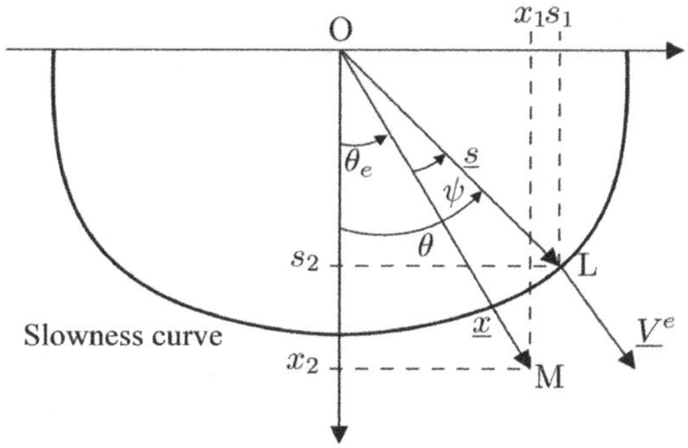

Figure 1.22. *Bulk wave generation in an anisotropic medium. Position of the stationary phase point L on the slowness curve, corresponding to the observation point M*

By introducing the wave number $k_J^e = k_J\cos\psi = \omega/V_J^e$, corresponding to the energy velocity V_J^e, we get:

$$\varphi_0 = k_J^e r \qquad \text{and} \qquad \varphi_0'' = -\frac{k_J^e r}{s_J^2\cos^2\theta^e}\frac{\mathrm{d}\theta^e}{\mathrm{d}\theta} \qquad [1.160]$$

The amplitude of the wave J transmitted in the direction θ is given by [1.132]:

$$b_J(r,\theta,\omega) = \omega F_k s_J\cos\theta^e E_{Jk}(s_J\sin\theta)\sqrt{\left|\frac{\mathrm{d}\theta}{\mathrm{d}\theta^e}\right|}\frac{\exp(ik_J^e r + i\varepsilon\pi/4)}{\sqrt{2\pi k_J^e r}} \qquad [1.161]$$

The phase $\varphi_0 = k_J^e r$ is proportional to the time r/V_J^e taken by the energy to reach the observation point distant from r. This result agrees with the fact that in a point source and point receiver configuration, the mechanical disturbance propagates at the energy velocity in the source–receiver direction.

In addition to the product $s_J(\theta)\cos\theta^e$, the angular dependence of the amplitude comes from the emission coefficient $E_{Jk}(s_J \sin\theta)$ and from the term $A = |\,\mathrm{d}\theta/\mathrm{d}\theta^e|$. This focusing factor, postulated by Maris (1971) in order to explain the phonon focusing observed in crystals, measures the anisotropy of the ray density, that is, the concentration of acoustic energy. For the displacement amplitude, the relevant quantity is $\sqrt{A}$. In highly anisotropic materials such as carbon-epoxy composites, this focusing effect dominates the variation with respect to the propagation direction of the emission coefficient (Corbel *et al.* 1993). Strong focusing ($A >> 1$) occurs when the direction of the energy velocity remains the same while the propagation direction varies ($|\,\mathrm{d}\theta^e/\mathrm{d}\theta| << 1$). Let us note that the stationary phase approximation breaks down for directions such that $\mathrm{d}\theta^e/\mathrm{d}\theta = 0$, for which $\varphi_0'' = 0$. The existence of a singularity for the displacement amplitude corresponds to points at which the curvature of the slowness curve vanishes, for example, in the presence of caustics. The ray model also becomes inaccurate when the emission factor $E_{Jk}(s_1)$ is canceled for, or in the vicinity of, the value s_1^0.

Therefore, the exact calculation of the dynamic response of an anisotropic solid half-space to a local and transient solicitation is complex. An analytical solution consists of determining the Green's function $G(s_1, x_2, p)$ in the space of the Laplace transform for time (complex variable p) and in the space of the Fourier transform according to x_1 (complex variable $k_1 = -ips_1$). The inversion of these transforms in order to calculate the impulse response $g(x_1, x_2, t)$, for example, by using the Cagniard-De Hoop method, requires rigorous mathematical developments (Poncelet and Deschamps 2009).

1.2.2.2. *Rayleigh wave*

On the surface of an elastic solid, the disturbance arising from a line source is composed of a part corresponding to the head wave and of a part corresponding to the Rayleigh wave (Figure 1.18). Unlike the head wave, the amplitude of the Rayleigh wave does not decrease during propagation. Then, far from the source, the Rayleigh wave predominates.

Relation [1.113], namely $u_i = Y_{ij}T_j$, shows that the components Y_{ij} of the admittance matrix $\boldsymbol{Y}$ are the generation coefficients of the displacements in the plane of the surface ($i = 1$) or out-of-plane ($i = 2$), produced by a uniform distribution of tangential mechanical traction ($j = 1$) or normal mechanical traction ($j = 2$), applied on the free surface of an elastic solid. These coefficients only depend on the slowness component $s_1 = k_1/\omega$. The displacement generated by a spatially varying force $T_j(x_1)$ is obtained by Fourier analysis:

$$U_i(x_1,\omega) = \frac{1}{2\pi}\int_{-\infty}^{+\infty} Y_{ij}(s_1)\hat{T}_j(k_1)e^{ik_1x_1}\,\mathrm{d}k_1 \qquad [1.162]$$

where $\hat{T}_j(k_1)$ is the Fourier transform of the spatial distribution $T_j(x_1)$. If the force is concentrated on the surface, along the line $x_1 = 0$: $T_j(x_1) = F_j\delta(x_1)$ and $\hat{T}_j(k_1) = F_j$. The mechanical displacement is expressed as:

$$U_i(x_1,\omega) = G_{ij}(x_1,\omega)F_j \quad \text{where} \quad G_{ij}(x_1,\omega) = \frac{\omega}{2\pi}\int_{-\infty}^{+\infty} Y_{ij}(s_1)e^{i\omega s_1x_1}\,\mathrm{d}s_1 \qquad [1.163]$$

is the spectral Green's function. The integral only depends on a single variable: the product ωx_1. For an isotropic solid, the Y_{ij} components take the form (Appendix 2):

$$Y_{ij}(s_1) = \frac{N_{ij}(s_1)}{i\omega\mu\Delta(s_1)} \qquad [1.164]$$

where $\Delta(s_1)$ is the Rayleigh determinant. The normal slownesses s_{2L} and s_{2T} are imaginary:

$$s_{2L} = i\sqrt{s_1^2 - s_L^2} \quad \text{and} \quad s_{2T} = i\sqrt{s_1^2 - s_T^2} \qquad [1.165]$$

so that the displacements of the Rayleigh wave ($e^{i\omega s_{2L}x_2}$ and $e^{i\omega s_{2T}x_2}$) decrease with depth $x_2 > 0$ in the material. The Rayleigh determinant [1.142] is written as:

$$\Delta(s_1) = \left(2s_1^2 - s_T^2\right)^2 - 4s_1^2\sqrt{\left(s_1^2 - s_T^2\right)\left(s_1^2 - s_L^2\right)} \qquad [1.166]$$

Each solution $s_1 = \pm s_R$ with $s_R > s_T > s_L$ to the Rayleigh equation $\Delta(s_1) = 0$ corresponds to a pole of all admittance coefficients Y_{ij}. The contribution from these poles to the integral [1.163] provides the Green's function G_{ij}^R for the surface displacement of the Rayleigh wave. This contribution is calculated by the residue method (Figure 1.23). The integration contour is closed by a semicircle $\mathcal{C}_+$ ($\mathcal{C}_-$) at infinity in the upper (lower) s_1 half-plane when x_1 is positive (negative), so that the factor $e^{i\omega s_1 x_1}$ is zero on each semicircle.

The poles $\pm s_R$ are located on the integration contour because we have neglected the attenuation. To take this effect into account, we must associate a positive imaginary part with the pole $+s_R$, corresponding to the propagation in the positive x_1 direction ($s_R \to s_R + i\alpha$), so that the factor $e^{-\omega\alpha x_1}$ is canceled when $x_1 \to +\infty$. Equivalently, a negative imaginary part must be associated with the pole $-s_R$ corresponding to the propagation in the negative x_1 direction ($-s_R \to -s_R - i\alpha$). The contribution of the positive pole (integrating anticlockwise around C_+) is given by:

$$G_{ij}(x_1, \omega) = \frac{1}{2i\pi\mu} \int_{-\infty}^{+\infty} \frac{N_{ij}(s_1)}{\Delta(s_1)} e^{i\omega s_1 x_1} \, \mathrm{d}s_1 = \frac{1}{\mu} \mathrm{Rs} \left[\frac{N_{ij}(s_1)}{\Delta(s_1)} e^{i\omega s_1 x_1} \right]_{s_R} \quad [1.167]$$

that is:

$$G_{ij}^{R+}(x_1, \omega) = \frac{N_{ij}(s_R)}{\mu(\mathrm{d}\Delta/\mathrm{d}s_1)_{s_R}} e^{i\omega s_R x_1} \quad \text{for} \quad x_1 > 0 \quad [1.168]$$

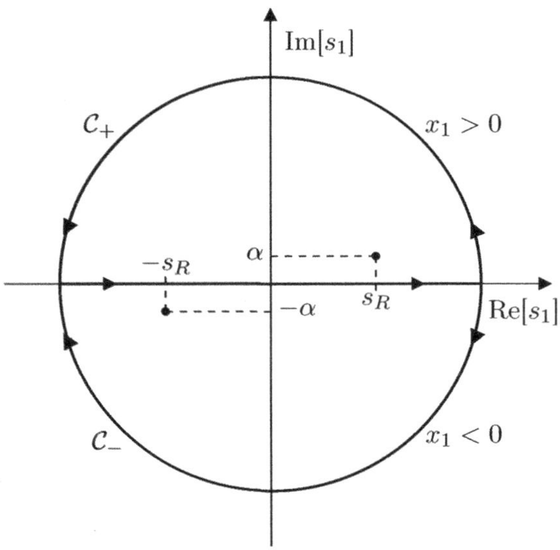

Figure 1.23. *Integration contour in the complex plane to evaluate the contribution of poles $\pm s_R$. In the upper half-plane (Im$[s_1] > 0$), the factor $\exp(i\omega s_1 x_1)$ approaches zero when the radius of the semicircle C_+ tends to infinity*

Using the property which states that $(\mathrm{d}\Delta/\mathrm{d}s_1)$ is an odd function s_1, the contribution of the negative pole (integrating clockwise around C_-) is:

$$G_{ij}^{R-}(x_1, \omega) = \frac{N_{ij}(-s_R)}{\mu(\mathrm{d}\Delta/\mathrm{d}s_1)_{s_R}} e^{-i\omega s_R x_1} \quad \text{for} \quad x_1 < 0 \quad [1.169]$$

The components of the matrix $\boldsymbol{N}$ (formulae [A2.27] in Appendix 2) are equal to:

$$N_{11}(s_R) = is_T^2\sqrt{s_R^2 - s_T^2} \quad \text{and} \quad N_{22}(s_R) = is_T^2\sqrt{s_R^2 - s_L^2} \qquad [1.170]$$

and:

$$N_{21}(s_R) = \frac{s_T^2}{2s_R}(2s_R^2 - s_T^2) = -N_{12}(s_R) \qquad [1.171]$$

The sign of the displacement depends on the parity of the numerator. Since N_{22} (N_{21}) is even (odd), a normal (tangential) force creates an equal (opposite) normal displacement moving off the source. The calculation (which is quite long) of the derivative of the Rayleigh determinant $\Delta(s_1)$ for $s_1 = s_R$ leads to an expression that depends on the parameters $\zeta = s_R^2/s_T^2$ and $\kappa = s_L/s_T$:

$$s_R\left(\frac{\mathrm{d}\Delta}{\mathrm{d}s_1}\right)_{s_R} = -2s_T^4\frac{8\zeta^3(1-\kappa^2) - 4\zeta + 1}{(2\zeta - 1)^2} \qquad [1.172]$$

Substituting the expressions for N_{12} and N_{22} in equation [1.168] provides the surface displacements (u_1 and u_2) of the Rayleigh wave generated by a line source of normal force density F_2 (Lamb 1904; Achenbach 1987). Similarly, substituting N_{11} and N_{21} provides the Rayleigh wave displacements (u_1 and u_2) generated by a line source of tangential force density F_1; for example, for $x_1 > 0$:

$$\begin{aligned} U_2 = G_{21}^{R^+}F_1 &= \frac{N_{21}(s_R)}{\mu(\mathrm{d}\Delta/\mathrm{d}s_1)_{s_R}}F_1e^{i\omega s_R x_1} \\ &= \frac{s_T^4(2\zeta - 1)}{2\mu s_R(\mathrm{d}\Delta/\mathrm{d}s_1)_{s_R}}F_1e^{i\omega s_R x_1} \end{aligned} \qquad [1.173]$$

Considering [1.172], the normal displacement is given by:

$$U_2 = b_{21}F_1e^{i\omega s_R x_1} \quad \text{with} \quad b_{21} = \frac{R}{4\mu} \quad \text{and} \quad R = \frac{(2\zeta-1)^3}{8\zeta^3(1-\kappa^2) - 4\zeta + 1} \qquad [1.174]$$

REMARK.–

– For $x_1 > 0$, the displacement u_2 is negative, because a positive tangential force F_1 compresses the material; in the ($x_1 < 0$) region, the same force creates a depression at the level of the free surface and u_2 is positive.

– The ratio of amplitudes $b_{21}/b_{11} = N_{21}(s_R)/N_{11}(s_R)$ is equal to that of transverse and longitudinal components of the Rayleigh wave (relation [3.23] in Volume 1): the disturbance propagates on the surface without deforming it.

– N_{21} is real, N_{11} is imaginary: the two displacement components are in phase quadrature.

ORDER OF MAGNITUDE.– For duralumin, $\zeta = 1.145$ and $\kappa = 0.486$, hence $R = 0.384$; with $\mu \approx 25$ GPa, the amplitude of the normal displacement of the Rayleigh wave created by a linear tangential force, of density $F_1 = 1$ N/cm, is $b_{21}F_1 = 0.38$ nm.

The time displacement, that is, the Fourier transform of $U_2(x_1, \omega)$:

$$
\begin{aligned}
u_2(x_1,t) &= \frac{1}{2\pi}\int_{-\infty}^{+\infty} U_2(x_1,\omega)e^{-i\omega t}\,\mathrm{d}\omega \\
&= \frac{b_{21}F_1}{2\pi}\int_{-\infty}^{+\infty} e^{-i\omega(t-s_R x_1)}\,\mathrm{d}\omega
\end{aligned}
\qquad [1.175]
$$

is a Dirac function:

$$
u_2(x_1,t) = b_{21}F_1\delta(t - s_R x_1) \qquad \text{for} \qquad x_1 > 0 \qquad [1.176]
$$

The mechanical displacement generated by a distribution of impulsive forces $F_1 q(t)$ is obtained through the convolution with $q(t)$:

$$
u_2(x_1,t) = b_{21}F_1 q(t - s_R x_1) \qquad [1.177]
$$

It has the same time-dependence $q(t)$ as the force that created it, while the temporal form of the tangential component is the Hilbert transform of $q(t)$ (Royer 2001).

In the case of a normal force of density F_2, since $N_{12}(s_R) = -N_{21}(s_R)$, the expression for the tangential component $U_1 = G_{12}^{R_+}F_2$ is deduced from equation [1.173], by changing the sign and by replacing F_1 by F_2. The tangential displacement reproduces the time variation $q(t)$ of the normal force:

$$
u_1(x_1,t) = -b_{21}F_2 q(t - s_R x_1) \qquad \text{with} \qquad b_{21} = \frac{R}{4\mu} \qquad [1.178]
$$

Then, it is the normal component which is the Hilbert transform of $q(t)$.

1.2.3. *Generation in a plate*

In practice, the propagation medium has finite dimensions. Let us study the generation in a plate, of thickness d, by a line source of normal forces. The observation point is located on the same side of the plate at a distance ℓ from the source. Assuming a Gaussian spatial distribution, uniform along x_3, of width a along x_1, the force density is of the form:

$$F_2(x_1, x_2 = 0, t) = F\frac{e^{-(x_1/a)^2}}{\sqrt{\pi}a}q(t) \qquad \text{with} \qquad q(t) = \frac{t}{\tau^2}e^{-t/\tau}H(t) \qquad [1.179]$$

The function $q(t)$ is normalized to unity and its duration at half-height is $\Delta t \approx 2.4\tau$. The width at -3 dB of its frequency spectrum $Q(\omega)$ is $\Delta f \approx 0.1/\tau$ (Figure 1.24):

$$\int_0^\infty q(t)\,\mathrm{d}t = 1 \qquad \text{and} \qquad Q(\omega) = \frac{1}{(1 + i\omega\tau)^2} \qquad [1.180]$$

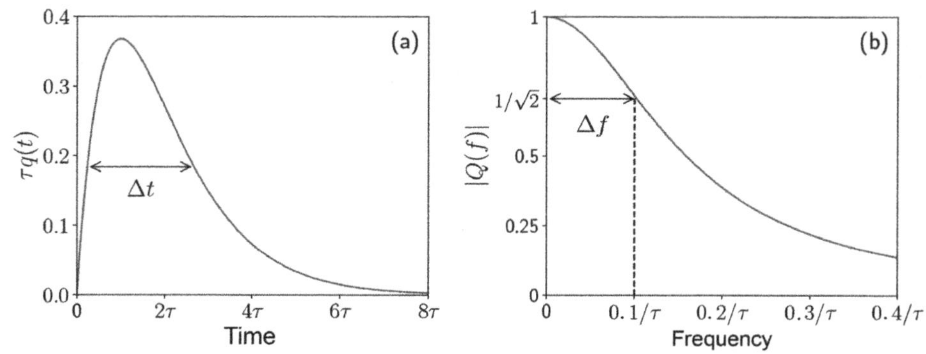

Figure 1.24. *Function $q(t)$ and modulus of its spectrum $Q(f)$. a) The pulse duration Δt is equal to 2.4τ; b) the width Δf of its spectrum is about $0.1/\tau$*

Compared to the case of a semi-infinite medium (section 1.2.1), the model takes into account all the upward waves (of amplitude a_I) and downward waves (of amplitude b_J). The expression [1.117] for the mechanical displacement is replaced by:

$$u_l = \sum_{I=1}^{3} a_I p_{lI} e^{i(k_1 x_1 + k_{2I} x_2 - \omega t)} + \sum_{J=1}^{3} b_J q_{lJ} e^{i(k_1 x_1 + k_{2J} x_2 - \omega t)} \qquad [1.181]$$

The normal components of the wave vector (k_{2I}, k_{2J}) and the polarizations (p_{lI}, q_{lJ}) are determined by solving the Christoffel equation. The amplitudes (a_I, b_J) of the partial waves are deduced from the boundary conditions on the two free surfaces of the plate. In the case of an isotropic solid, the sums are limited to two terms, with $I, J = 1$ or 2 for longitudinal or transverse waves, respectively.

Figure 1.25 shows the mechanical displacement generated in a stainless steel plate of thickness $d = 4$ mm, for $\tau = 10$ ns and a force $F = 10^{-2}$ N, that is, $F/a = 100$ N/m, when $a = 0.1$ mm. The tangential component u_1 and normal component u_2 were calculated at a distance $\ell = 6$ mm on the same face as the source. For both components, the first signal H corresponds to the head wave that propagates on the surface at a velocity $V_L = 5\ 634$ m/s. The largest displacement R is that of the Rayleigh wave that propagates at the velocity $V_R = 3\ 015$ m/s. The tangential component of the Rayleigh wave (Figure 1.25(a)) reproduces the time variation $q(t)$ of the force while the normal component (Figure 1.25(b)) is similar to the Hilbert transform of $q(t)$.

With a period $\tau = 10$ ns, that is, $\Delta f = 10$ MHz, and for a frequency $f = 5$ MHz, the wavelength $\lambda_R = V_R/f \approx 0.6$ mm is much smaller than the plate thickness. The Rayleigh wave propagates as if the medium were semi-infinite, ignoring the second face that is at a distance greater than $2\lambda_R$.

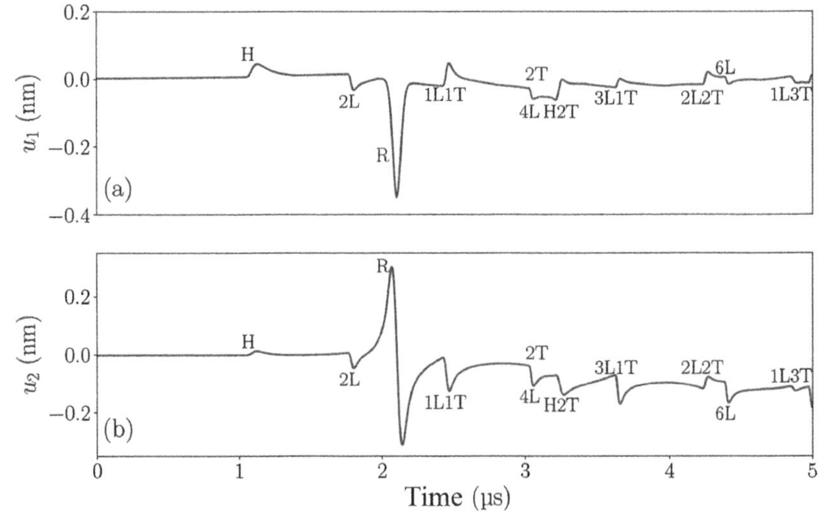

Figure 1.25. *a) Tangential displacement u_1 and b) normal displacement u_2, created by a line of normal forces in a 4-mm thick steel plate. The observation point is located on the same face, at a distance $\ell = 6$ mm from the source*

The other echoes are due to the successive reflections with the possible conversion of bulk waves on each free surface (section 2.4.1, Volume 1). For example, the echo 1T1L corresponds to a transverse wave on the forward path and a longitudinal wave on the return path converted during the reflection on the opposite face (Figure 1.26). Their arrival times are predicted by a ray model, using the relation:

$$t_{nLmT} = d\left(\frac{n}{V_L \cos\theta_L} + \frac{m}{V_T \cos\theta_T}\right) \qquad [1.182]$$

where integers n and m, whose sum is necessarily even, correspond to the number of single paths from one face to another for L and T waves, between the line source and the receiver. For equal times of flight (for example, when $n = m$: $t_{nTnL} = t_{nLnT}$), only one of these paths is shown in Figure 1.25.

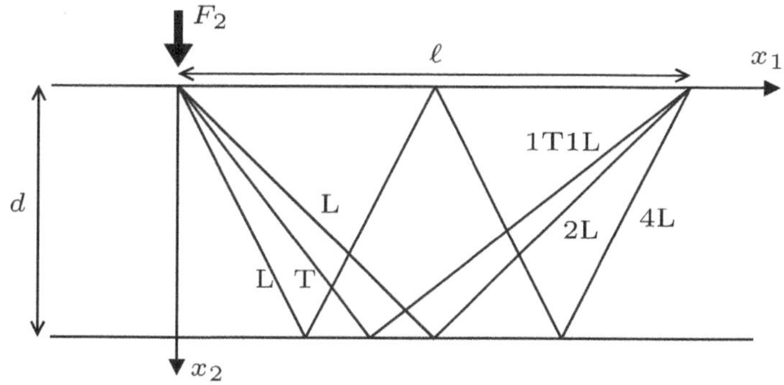

Figure 1.26. *Generation in a plate by a normal force. First echoes received at the point $x_1 = \ell$*

The angles of incidence θ_L and θ_T, with respect to the normal to the plate, depend on the number of single paths for each wave. The projection of these paths on the x_1 axis is:

$$\ell = d\,(n\tan\theta_L + m\tan\theta_T) \qquad \text{with} \qquad n + m = 2p \qquad [1.183]$$

Since angles θ_L and θ_T are related by the Snell–Descartes law $\sin\theta_T = \kappa\sin\theta_L$, equation [1.183] has only one unknown, for example, $\sin\theta_L$. Once the angles of incidence are determined, the arrival time t_{nLmT} can be calculated using formula [1.182]. In Figure 1.25, the first echoes are indicated at their arrival times.

The echo H2T is different: it arises from the radiation of the head wave into a transverse wave at the critical angle $\theta_c^L = \arcsin(\kappa)$. This wave is then reflected off the rear face of the plate, before reaching the receiver. The transit time associated with this path is given by:

$$t_{H2T} = \frac{\ell - 2d\tan\theta_c^L}{V_L} + \frac{2d}{V_T\cos\theta_c^L} \qquad [1.184]$$

Due to the energy loss through radiation toward the core of the plate, the head wave H and the echo H2T are rapidly attenuated when the distance ℓ increases.

In the case of a thin plate, whose thickness d is no longer large with respect to the wavelengths λ_R, λ_T and λ_L, the different echoes are superimposed. Far from the source, the successive reflections of L and T waves give rise to eigenmodes guided by the plate. For example, when d is of the order of λ_R, the coupling (symmetrical or antisymmetrical) of the Rayleigh waves propagating on both faces of the plate is at the origin of the two fundamental modes S_0 and A_0. These Lamb waves, studied in section 4.3.1 in Volume 1, are composed of both longitudinal and transverse vertical displacements. The proportion of these components and, thus, the phase velocity of Lamb waves vary with frequency.

1.3. Radiation of elementary spherical sources

In this section, we calculate the mechanical displacement radiated in an isotropic medium by a moving sphere. This sphere can expand and contract along its radius, or oscillate in a rotational or translational motion.

Let R_s be the reference frame with spherical coordinates (r, θ, φ) and center O. The Cartesian coordinates are then defined in terms of spherical coordinates through relations [A1.21]. The displacement field $\underline{u}$, which is a solution to the propagation equation, is searched using the Helmholtz decomposition:

$$\underline{u} = \underline{\text{grad}}\,\phi + \underline{\text{rot}}\,\underline{\psi} \qquad [1.185]$$

with:

$$\phi = \phi_L \quad \text{and} \quad \underline{\psi} = \underline{\text{rot}}\,(r\phi_T\underline{e}_r) + \underline{\text{rot}}\left[\underline{\text{rot}}\left(\frac{r}{k_T}\phi_S\underline{e}_r\right)\right] \qquad [1.186]$$

where ϕ_L, ϕ_T and ϕ_S are scalar potentials, solutions to the Helmholtz equations:

$$\Delta\phi_L + k_L^2\phi_L = 0 \quad \text{and} \quad \Delta\phi_M + k_T^2\phi_M = 0 \qquad \text{with M = T, S} \qquad [1.187]$$

By assuming that the motions are independent of the angle φ, the components of the displacement field are expressed in spherical coordinates as:

$$\begin{cases} u_r(r,\theta) = \dfrac{\partial \phi_L}{\partial r} - \dfrac{1}{r\sin\theta}\dfrac{\partial}{\partial\theta}\left(\sin\theta\dfrac{\partial\phi_T}{\partial\theta}\right) \\ u_\theta(r,\theta) = \dfrac{1}{r}\dfrac{\partial\phi_L}{\partial\theta} + \dfrac{1}{r}\dfrac{\partial}{\partial\theta}\dfrac{\partial(r\phi_T)}{\partial r} \\ u_\varphi(r,\theta) = -k_T\dfrac{\partial\phi_S}{\partial\theta} \end{cases} \quad [1.188]$$

1.3.1. *Pulsating sphere*

Let us consider a sphere of radius R animated by a purely radial displacement (pulsating sphere). When the three potentials ϕ_L, ϕ_T and ϕ_S do not depend on the angle θ, only the radial displacement is non-zero; it derives from the scalar potential ϕ_L associated with the longitudinal wave:

$$u_r(r,\theta,\varphi) = \frac{\partial\phi_L}{\partial r} \quad [1.189]$$

This is the solution to the equation:

$$\frac{\partial^2\phi_L}{\partial r^2} + \frac{2}{r}\frac{\partial\phi_L}{\partial r} + k_L^2\phi_L = 0 \quad [1.190]$$

hence:

$$\phi_L(r) = \frac{1}{k_L^2}\left[\frac{a_L}{r}e^{ik_Lr} + \frac{b_L}{r}e^{-ik_Lr}\right] \quad [1.191]$$

The first term represents a divergent wave, the second one a convergent wave, with amplitudes a_L and b_L. In the surrounding medium ($r > R$), the radial displacement u_r and radial stress σ_{rr} of the divergent wave are written as (formula [A1.17] in Volume 1):

$$\begin{cases} u_r(r) = \dfrac{a_L}{(k_Lr)^2}(ik_Lr - 1)e^{ik_Lr} \\ \sigma_{rr}(r) = -\dfrac{a_L}{k_L^2r^3}\left[(\lambda+2\mu)(k_Lr)^2 + 4\mu(ik_Lr-1)\right]e^{ik_Lr} \end{cases} \quad [1.192]$$

In the following, two cases are examined. If the sphere is animated by a radial displacement of amplitude U, the continuity of the displacement $u_r(R) = U$ leads to the following expression for the amplitude:

$$a_L = U\frac{(k_LR)^2}{ik_LR - 1}e^{-ik_LR} \qquad [1.193]$$

or again, if $\tilde{k}_L = k_LR << 1$:

$$a_L = -\tilde{k}_L^2 U \qquad [1.194]$$

If the sphere is a cavity with a pressure p_0 that is assumed to be uniform on its surface, the continuity of the radial stress $\sigma_{rr}(R) = -p_0$ allows us to express the amplitude a_L through the relation:

$$a_L = \frac{p_0\tilde{k}_L^2 Re^{-i\tilde{k}_L}}{(\lambda + 2\mu)\tilde{k}_L^2 + 4\mu(i\tilde{k}_L - 1)} \qquad [1.195]$$

or again, if $\tilde{k}_L << 1$:

$$a_L = -\frac{p_0\tilde{k}_L^2 R}{4\mu} \qquad [1.196]$$

1.3.2. *Oscillating sphere with a rotational motion*

Let there be a sphere animated by a rotational motion around the z axis, of angular amplitude Ω:

$$u_\varphi(R, \theta) = \Omega R \sin\theta \qquad [1.197]$$

Since the movement of the sphere is, on the one hand, independent of the angle φ and, on the other hand, only polarized along the $\underline{e}_\varphi$ axis, the wave excited by the rotation of the sphere derives only from the potential ϕ_S:

$$u_\varphi(r, \theta) = -k_T\frac{\partial\phi_S}{\partial\theta} \quad \text{for} \quad r > R \qquad [1.198]$$

The displacement of the sphere imposes the following form for the potential ϕ_S:

$$\phi_S(r, \theta) = \frac{f_S(r)}{k_T}\cos\theta \qquad [1.199]$$

where the function f_S satisfies the Helmholtz equation [A1.22]:

$$\frac{\partial^2 f_S}{\partial r^2} + \frac{2}{r}\frac{\partial f_S}{\partial r} - \frac{2}{r^2} f_S + k_T^2 f_S = 0 \qquad [1.200]$$

The solution is a linear combination of first-order spherical Hankel functions of the first and second kinds, $h_1^{(1)}$ and $h_1^{(2)}$, that is:

$$f_S(r) = a_S h_1^{(1)}(k_T r) + b_S h_1^{(2)}(k_T r) \qquad [1.201]$$

These functions are associated with divergent and convergent waves. Only the divergent wave is generated by the motion of the sphere. With $b_S = 0$, the displacement is given by:

$$u_\varphi(r,\theta) = a_S h_1^{(1)}(k_T r) \sin\theta \qquad [1.202]$$

The continuity of the displacement u_φ on the surface of the sphere [1.197] imposes:

$$a_S = \frac{R\Omega}{h_1^{(1)}(\tilde{k}_T)} \qquad [1.203]$$

At low frequency and in the far field ($k_T r >> 1$ and $\tilde{k}_T = k_T R << 1$), the displacement is given by:

$$u_\varphi(r,\theta) = -ik_T R^3 \Omega \frac{e^{ik_T r}}{r} \sin\theta \qquad [1.204]$$

1.3.3. *Oscillating sphere with a translational motion*

Let there be a sphere animated by a translational motion of amplitude U along the z axis. In the reference frame with spherical coordinates, the displacement components at each point on the sphere are expressed by:

$$u_r(R,\theta) = U\cos\theta, \qquad u_\theta(R,\theta) = -U\sin\theta \qquad \text{and} \qquad u_\varphi(R,\theta) = 0 \qquad [1.205]$$

Since the motion of the sphere is independent of the angle φ and is contained in the $(\underline{e}_r, \underline{e}_\theta)$ plane, the wave emitted by the sphere is described by potentials ϕ_L and ϕ_T of the type:

$$\phi_M(r,\theta) = f_M(r)\cos\theta \qquad \text{with} \qquad \text{M} = \text{L, T} \qquad [1.206]$$

The functions $f_M(r)$ are the solutions to equation [1.200]. Retaining only terms that correspond to divergent waves, they take the form:

$$f_M(r) = \frac{a_M}{k_M} h_1^{(1)}(k_M r) \qquad \text{with} \qquad \mathrm{M = L, T} \tag{1.207}$$

and the non-zero components of the displacement are given by:

$$\begin{cases} u_r(r,\theta) = \left[a_L \dfrac{\partial h_1^{(1)}(k_L r)}{\partial (k_L r)} + \dfrac{2a_T}{k_T r} h_1^{(1)}(k_T r) \right] \cos\theta \\ u_\theta(r,\theta) = -\left[\dfrac{a_L}{k_L r} h_1^{(1)}(k_L r) + \dfrac{a_T}{k_T r} \dfrac{\partial [r h_1^{(1)}(k_T r)]}{\partial r} \right] \sin\theta \end{cases} \tag{1.208}$$

Considering expression [A1.39], the application of the boundary conditions at $r = R$ leads to the following relations, by writing $\tilde{k}_L = k_L R$ and $\tilde{k}_T = k_T R$:

$$\begin{cases} \dfrac{a_L}{\tilde{k}_L} \left[h_1^{(1)}(\tilde{k}_L) - \tilde{k}_L h_2^{(1)}(\tilde{k}_L) \right] + \dfrac{2a_T}{\tilde{k}_T} h_1^{(1)}(\tilde{k}_T) = U \\ \dfrac{a_L}{\tilde{k}_L} h_1^{(1)}(\tilde{k}_L) + \dfrac{a_T}{\tilde{k}_T} \left[2h_1^{(1)}(\tilde{k}_T) - \tilde{k}_T h_2^{(1)}(\tilde{k}_T) \right] = U \end{cases} \tag{1.209}$$

The resolution of these equations leads to the expressions:

$$\begin{cases} a_L = -\dfrac{\tilde{k}_L \tilde{k}_T h_2^{(1)}(\tilde{k}_T) U}{\tilde{k}_L h_2^{(1)}(\tilde{k}_L) \left[\tilde{k}_T h_2^{(1)}(\tilde{k}_T) - 2h_1^{(1)}(\tilde{k}_T) \right] - \tilde{k}_T h_2^{(1)}(\tilde{k}_T) h_1^{(1)}(\tilde{k}_L)} \\ a_T = -\dfrac{\tilde{k}_L \tilde{k}_T h_2^{(1)}(\tilde{k}_L) U}{\tilde{k}_L h_2^{(1)}(\tilde{k}_L) \left[\tilde{k}_T h_2^{(1)}(\tilde{k}_T) - 2h_1^{(1)}(\tilde{k}_T) \right] - \tilde{k}_T h_2^{(1)}(\tilde{k}_T) h_1^{(1)}(\tilde{k}_L)} \end{cases} \tag{1.210}$$

that is, when the longitudinal and transverse wavelengths are large compared to the radius R of the sphere ($\tilde{k}_M << 1$):

$$a_M = \frac{iU k_L k_T}{k_L^2 + 2k_T^2} \tilde{k}_M e^{-i\tilde{k}_M} \qquad \text{with} \qquad \mathrm{M = L, T} \tag{1.211}$$

Given the limit of the function $h_1^{(1)}$ for large arguments (formula [A1.40]), the approximate expressions for the displacement components in the far field are:

$$u_r(r,\theta) = -i a_L \frac{e^{ik_L r}}{r} \cos\theta \qquad \text{and} \qquad u_\theta(r,\theta) = i a_T \frac{e^{ik_T r}}{r} \sin\theta \tag{1.212}$$

The mechanical displacement is calculated in the far field for a bead buried in an epoxy matrix. The components u_r and u_θ, associated, respectively, with the longitudinal and transverse waves, are plotted as functions of the angle θ in Figure 1.27 for the case of a sphere whose radius is (a) small compared to the wavelength, and (b) large compared to the wavelength. These diagrams depict the directivity of the mechanical displacement. For a sphere radius small compared to the wavelength, the radiation of the transverse wave is more intense than that of the longitudinal wave, while they are comparable for a sphere radius larger than the wavelengths. In both cases, the oscillation of the sphere in the z direction leads to the radiation of a longitudinal wave in this direction and of a transverse wave in the xy plane: the sphere motion imposes the polarization of the emitted waves.

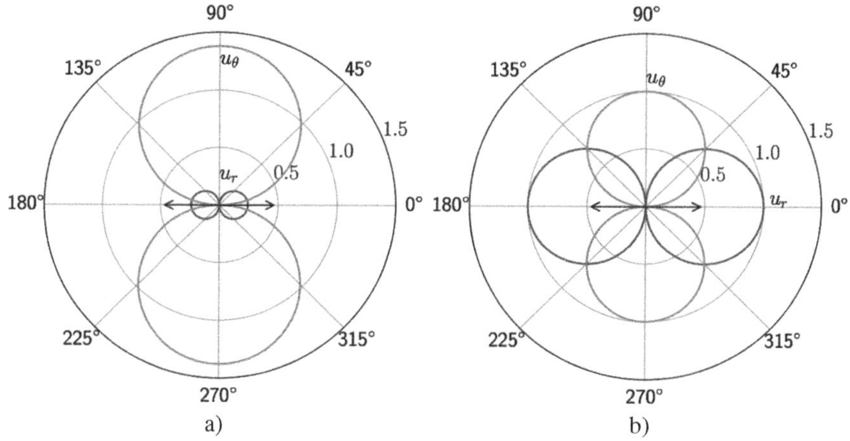

Figure 1.27. *Sphere embedded in an epoxy resin and oscillating in translation along the z axis. Components u_r and u_θ of the mechanical displacement as functions of the angle θ when a) $\lambda >> R$ and b) $\lambda << R$. For a color version of this figure, see www.iste.co.uk/royer/waves2.zip*

2

Scattering of Elastic Waves

In Volume 1, the propagation medium, assumed to be homogeneous, was either infinite (Chapter 1), divided into two parts by a plane interface (Chapters 2 and 3), or limited in one or two dimensions (Chapter 4), the waves propagating freely in this or other dimensions. Here, the propagation medium is locally inhomogeneous; these heterogeneities may be inherent to the structure or may result from manufacturing defects (inclusions) or from the aging of the parts (cracks). Waves scattered by an obstacle propagate away from it, possibly in privileged directions, even if the medium is isotropic. Several applications in underwater acoustics are based on target scattering: detecting mines or ships by an active sonar, locating shoals of fish, or mapping ocean floors. In non-destructive testing by ultrasound, the information carried by the scattered waves is used to characterize these heterogeneities or to prevent damage, such as the delamination of composite materials. Scatterers can be deliberately introduced, for instance, to attenuate backward waves emitted by a piezoelectric transducer and thus dampen its responses (section 3.1). Recently, the development of periodic structures (phononic crystals) or heterostructures made of resonant inclusions dispersed in a matrix (acoustic metamaterials) opened the way to the filtering and the control of elastic wave propagation.

The interaction of a (plane) elastic wave with a particle embedded in a matrix gives rise to many phenomena (Figure 2.1), such as:

– the reflection at the particle–matrix interface;

– the transmission inside the particle with a possible mode conversion, and then internal reflections and refraction;

– the conversion into waves propagating at the interface with the matrix, and then re-emission into bulk waves;

– the diffraction by the edges of the object;

– the absorption of part of the energy within the particle.

All of these phenomena, which can give rise to resonances involving bulk, surface or guided waves, have been studied separately in Volume 1. The scattering theory accounts for these phenomena in a global way. The approach consists of defining, in the harmonic case, an incident field and a scattered field that satisfy the propagation equations in each of the two media. The boundary conditions at the interface between the scatterer and the matrix are applied to the total field.

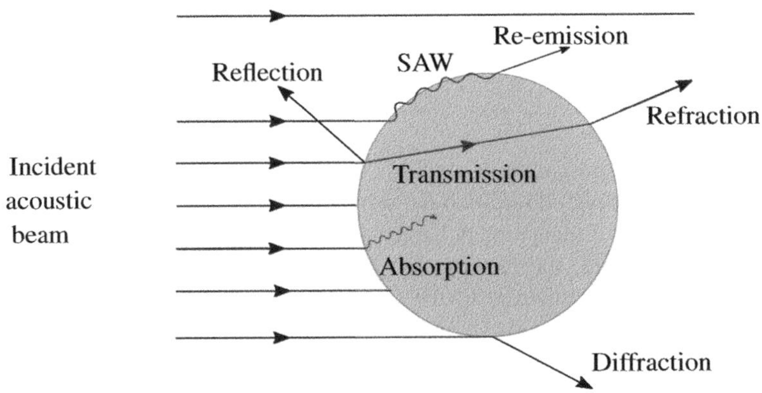

Figure 2.1. *Interaction of an elastic wave with a scatterer. The incident energy is either reflected, or transmitted and then refracted, or converted into a surface acoustic wave (SAW) and then re-emitted, diffracted by the edges of the scatterer, or partially absorbed inside the scatterer*

Formally, the displacement field $\underline{u}$ is written as the superposition of an incident displacement field $\underline{u}_i$ and a scattered field $\underline{u}_d$, composed *a priori* of all the propagation modes M:

$$\underline{u} = \underline{u}_i + \underline{u}_d \qquad \text{with} \qquad \underline{u}_d = \sum_M \underline{u}_M^d \qquad [2.1]$$

In a perfect fluid or an isotropic solid, the acoustic field associated with longitudinal waves is determined through the knowledge of a scalar potential Φ, while the field associated with transverse waves in the solid derives from a vector potential $\underline{\psi}$. For obstacles having a canonical form (cylinder or sphere), the method of analysis consists of:

– expanding the potentials of the incident wave on the basis of harmonic solutions (cylindrical or spherical) to the Helmholtz equation (see Appendix 1);

– postulating a similar development, with unknown coefficients, for the scattered waves;

– determining these *scattering coefficients* by applying, for both fields, the boundary conditions at the fluid–solid interface.

Due to the orthogonality of cylindrical and spherical harmonics, these conditions can be applied independently to each harmonic of order n. The problem can be reduced to the resolution of a linear system having, *a priori*, an infinite number of equations. Finally, the scattered field is obtained as an expansion of cylindrical or spherical harmonics. Fortunately, the number of terms can be reduced according to the convergence of the series of harmonic functions.

The interest in this model is to predict the evolution of the acoustic intensity scattered in each direction, according to the elastic and geometrical characteristics of the scatterer. The complexity of the calculation will quickly convince the reader of the advantage of considering the symmetry of the "incident wave + scatterer" configuration in order to reduce the number of modes to take into account in the developments. From this point of view, it is very simple to use the principle stated by Pierre Curie in 1894: "When certain causes produce certain effects, the elements of symmetry of the causes must be found in the effects produced."

As an illustration, let us study two cases that will be examined in sections 2.1 and 2.3.2.

1) A longitudinal plane wave propagating along the x axis encounters a cylinder infinitely long along z (Figure 2.2(a)). The configuration is symmetric with respect to the xy plane: the scattered waves (longitudinal or transverse) are necessarily polarized in the xy plane: $\underline{u} = \underline{u}(r, \theta)$ with $u_z = 0$.

2) A longitudinal plane wave propagating along the z axis encounters a sphere (Figure 2.2(b)). The configuration is axisymmetric around the z axis: the scattered field is independent of the polar angle φ: $\underline{u} = \underline{u}(r, \theta)$ with $u_\varphi = 0$.

SCATTERING CROSS-SECTION.– One of the results of this analysis is the existence of three scattering regimes depending on the value of the wavelength with respect to the characteristic dimensions of the heterogeneities:

– the *Rayleigh scattering* regime, when the wavelength is large compared to the size of the hetereogeneity;

– the *resonant* regime, when the wavelength is comparable to the size of the hetereogeneity;

– the *geometric* regime, when the wavelength is small compared to the size of the hetereogeneity.

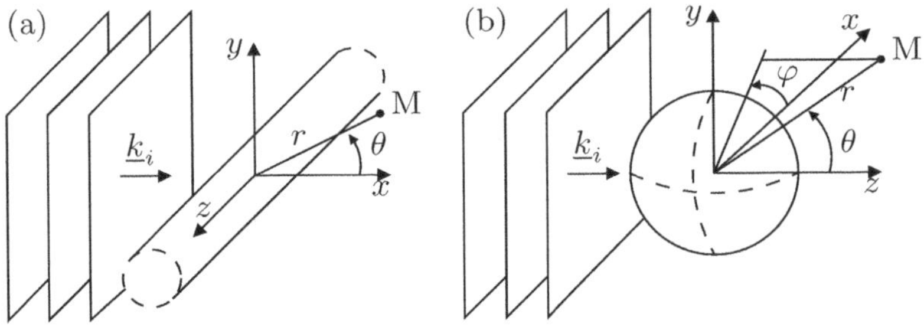

Figure 2.2. *Scattering of a longitudinal wave a) by a cylinder and b) by a sphere*

These regimes are characterized by a different dependence of the scattered power (as a function of frequency) and acoustic intensity (depending on the direction). These variations are expressed by two scattering cross-sections, which we define first in the case of a sphere. Using I_{inc} to denote the acoustic intensity (power per unit area) transported by the incident wave, the *scattering cross-section* σ is such that the product σI_{inc} is equal to the total power P_d scattered throughout the space.

This one is obtained by integrating the scattered intensity $I_d(r, \theta, \varphi)$ on a sphere, of very large radius r, surrounding the particle:

$$\sigma I_{inc} = P_d = \int_0^{2\pi} \int_0^{\pi} I_d(r, \theta, \varphi) r^2 \sin\theta \, \mathrm{d}\theta \, \mathrm{d}\varphi \qquad \text{with} \qquad r \to \infty \qquad [2.2]$$

σ is homogeneous to a surface and independent of r, because, in the far field, the intensity of the scattered waves decreases as $1/r^2$. This quantity is usually normalized with respect to the dimensions of the scatterer. In the case of a rigid (in acoustics) or opaque (in optics) sphere, whose radius R is very large compared to the wavelength λ, and for a directional beam, it is natural to assume that, in the absence of absorption, the scattered power corresponds to the rays intercepted by the particle, that is, $\pi R^2 I_{inc}$. Indeed, since σ is calculated at infinity, any ray, even a very slightly deflected one, leaves the geometric shadow cast by the obstacle. This is the case of waves diffracted by the disk of radius R, whose power is also equal to $\pi R^2 I_{inc}$. Indeed, we will see that, when kR approaches infinity, the scattering theory leads to a cross-section twice that predicted by the purely geometrical approach.

When the radius R is of the order of the wavelength, resonances and/or interference between the incident wave and the forward scattered waves considerably increase the efficiency of the scattering. This regime, described by the Mie theory in optics, is called *resonant*. Let us note that in acoustics the phenomena are more complex, especially in the case of an elastic scatterer buried in a solid matrix.

In order to characterize the scattering anisotropy, it is necessary to introduce a differential scattering *section* $\sigma_d(\theta, \varphi)$, such that:

$$\sigma = \int_0^{2\pi} \int_0^{\pi} \sigma_d(\theta, \varphi) \sin\theta \, \mathrm{d}\theta \, \mathrm{d}\varphi \quad \text{hence} \quad \sigma_d(\theta, \varphi) = \lim_{r \to \infty} r^2 \frac{I_d(r, \theta, \varphi)}{I_{inc}} \qquad [2.3]$$

Given the relation [2.2] and the $1/r^2$ decay of the scattered intensity, σ_d only depends on the direction defined by the polar angles θ and φ.

In the case of a *cylinder, infinite and homogeneous* along the z axis, the scattered acoustic intensity $I_d(r, \theta)$ decreases as $1/r$ in the far field. The previous relations are replaced by:

$$\sigma I_{inc} = \int_0^L \int_{-\pi}^{\pi} I_d(r, \theta) r \, \mathrm{d}\theta \, \mathrm{d}z = I_{inc} \int_{-\pi}^{\pi} \sigma_d(\theta) \, \mathrm{d}\theta \qquad [2.4]$$

where:

$$\frac{\sigma_d(\theta)}{L} = \lim_{r \to \infty} r \frac{I_d(r, \theta)}{I_{inc}} \qquad [2.5]$$

is the differential cross-section per unit length of the cylinder.

In the first section, entitled *acoustic scattering*, the scatterer is a cylinder immersed in a perfect fluid, therefore, the incident wave is longitudinal. Two cases are studied: a rigid cylinder and an elastic cylinder. In the second section, the elastic scattering by a cylinder embedded in an isotropic solid matrix is illustrated by a practical example for a longitudinal and a transverse incident wave. The scattering by a rigid or an elastic sphere buried in an isotropic solid matrix is treated in the third section. For both types of incident waves, the scattering cross-sections are calculated and their variations are illustrated in typical configurations. The phenomena of multiple scattering by a set of particles are discussed at the end of the chapter, using only a scalar approach.

2.1. Acoustic scattering by an immersed cylinder

In this section, we study the scattering of an acoustic wave by an infinitely long cylinder, of radius R, immersed in a fluid. The propagation is analyzed for time-harmonic waves of angular frequency ω and the time dependence in $e^{-i\omega t}$ is deliberately omitted. Let R_c be the cylindrical coordinate system (r, θ, z) of Appendix 1. The Cartesian coordinates are then defined in terms of cylindrical coordinates by relations [A1.1]. The field, contained in the xy plane, is independent of the z coordinate. The incident plane wave propagates in the x direction in a fluid of velocity V_f and mass density ρ_f. The velocity potential Φ_i associated with the incident wave, of amplitude a_f, is expanded on the cylindrical harmonics [A1.20]:

$$\Phi_i(r,\theta) = \frac{ia_f}{k_f} e^{ik_f x} = \frac{ia_f}{k_f} \sum_{n=0}^{\infty} \epsilon_n i^n J_n(k_f r) \cos(n\theta) \tag{2.6}$$

where J_n is the Bessel function of the first kind and where $\epsilon_n = 1$ for $n = 0$ and $\epsilon_n = 2$ for $n > 0$. The velocity potential of the wave scattered by the cylinder is also expanded on the cylindrical harmonics:

$$\Phi_d(r,\theta) = \frac{ia_f}{k_f} \sum_{n=0}^{\infty} \epsilon_n i^n T_n H_n(k_f r) \cos(n\theta) \tag{2.7}$$

where $H_n \equiv H_n^{(1)}$ is the Hankel function of the first kind, associated with a divergent wave. The amplitudes T_n of the cylindrical harmonics are to be determined from the boundary conditions. The total pressure and the radial velocity in the fluid are deduced from formulae [1.3] and [1.4]:

$$p_a(r,\theta) = a_f Z_f \sum_{n=0}^{\infty} \epsilon_n i^n \left[J_n(k_f r) + T_n H_n(k_f r)\right] \cos(n\theta) \tag{2.8}$$

and:

$$v_r(r,\theta) = -ia_f \sum_{n=0}^{\infty} \epsilon_n i^n \left[J_n'(k_f r) + T_n H_n'(k_f r)\right] \cos(n\theta) \tag{2.9}$$

where the derivatives J_n' and H_n' of the Bessel and Hankel functions are given by relations [A1.14].

The separation of variables method thus leads to express the acoustic fields in the form of modal developments. These developments are convergent series. Apart from the monopolar mode ($n = 0$) and the dipolar mode ($n = 1$), the contribution of the nth order mode to this series is preponderant, as long as n is less than the reduced wave number kR. For numerical reasons, the series may therefore be truncated and the number of modes to consider for a satisfactory convergence depends on the frequency domain. At low frequencies, that is, when $kR < 1$, only the monopolar and dipolar modes have a significant influence. It is, therefore, usual to retain only these two modes, which leads to simpler analytical developments.

2.1.1. *Scattering by a rigid cylinder*

If the cylinder is immobile and rigid, the radial component of the particle velocity cancels on its surface: $v_r(r = R, \theta) = 0$. Taking into account the expression [2.9], this condition leads to the relation:

$$\sum_{n=0}^{\infty} \epsilon_n i^n \left[J_n'(\tilde{k}_f) + T_n H_n'(\tilde{k}_f) \right] \cos(n\theta) = 0 \quad \text{with} \quad \tilde{k}_f = k_f R \qquad [2.10]$$

Given the orthogonality of the $\cos(n\theta)$ functions, for each cylindrical harmonic we get:

$$J_n'(\tilde{k}_f) + T_n H_n'(\tilde{k}_f) = 0 \qquad [2.11]$$

Given the expression [A1.14] for the derivative of Bessel and Hankel functions, the scattering coefficients T_n can be written as:

$$T_n = -\frac{J_n'(\tilde{k}_f)}{H_n'(\tilde{k}_f)} = -\frac{nJ_n(\tilde{k}_f) - \tilde{k}_f J_{n+1}(\tilde{k}_f)}{nH_n(\tilde{k}_f) - \tilde{k}_f H_{n+1}(\tilde{k}_f)} \qquad [2.12]$$

In order to calculate the scattering cross-section, the solutions must be expressed in the far field. Since the Hankel function of the first kind and its derivative have the following asymptotic behavior:

$$\lim_{x \to \infty} H_n(x) = \frac{1-i}{i^n} \frac{e^{ix}}{\sqrt{\pi x}} \quad \text{and} \quad \lim_{x \to \infty} H_n'(x) = \frac{1+i}{i^n} \frac{e^{ix}}{\sqrt{\pi x}} \qquad [2.13]$$

the radial velocity and acoustic pressure associated with the scattered wave are given by the approximate expressions:

$$\begin{cases} v_r^d(r,\theta) = a_f \dfrac{e^{ik_f r}}{\sqrt{r}} f(\theta) \\ p_a^d(r,\theta) = Z_f v_r^d(r,\theta) \end{cases} \quad \text{if} \quad k_f r >> 1 \qquad [2.14]$$

where the *far-field scattering function* $f(\theta)$ is equal to:

$$f(\theta) = \frac{1-i}{\sqrt{\pi k_f}} \sum_{n=0}^{\infty} \epsilon_n T_n \cos(n\theta) \qquad [2.15]$$

The scattered acoustic intensity is the time-average of the radial component of the Poynting vector:

$$I_d = \langle \underline{J} \cdot \underline{e}_r \rangle = \frac{1}{2} \operatorname{Re} \left[p_a^d \left(v_r^d \right)^* \right] \qquad [2.16]$$

In the far-field, I_d is proportional to the intensity of the incident plane wave I_{inc}:

$$I_d(r,\theta) = \frac{1}{2} Z_f \left|v_r^d\right|^2 = I_{inc} \frac{\sigma_d(\theta)}{r} \qquad \text{with} \qquad I_{inc} = \frac{1}{2} Z_f |a_f|^2 \tag{2.17}$$

and where:

$$\sigma_d(\theta) = |f(\theta)|^2 \tag{2.18}$$

is the *differential scattering cross-section*, defined by equation [2.5]. The scattering cross-section corresponds to the average of $\sigma_d(\theta)$ on a cylindrical surface of very large radius r:

$$\sigma = \int_0^1 \int_0^{2\pi} \sigma_d(\theta)\, \mathrm{d}\theta\, \mathrm{d}z = \int_0^{2\pi} |f(\theta)|^2\, \mathrm{d}\theta \tag{2.19}$$

Since the cylinder is uniform and infinitely long, the integration along z is limited to the unit length. Upon normalizing with respect to the geometric cross-section $2R$ of the cylinder of unit length, the substitution of the expression [2.15] of the far-field scattering function $f(\theta)$ into relation [2.19] leads to:

$$\gamma = \frac{\sigma}{2R} = \frac{2}{\tilde{k}_f} \sum_{n=0}^{\infty} \epsilon_n |T_n|^2 \qquad \text{with} \qquad \tilde{k}_f = k_f R \tag{2.20}$$

In the long wavelength regime, that is for $\tilde{k}_f << 1$, the summation appearing in the far-field scattering function $f(\theta)$ can be restricted to the first two modes. In this case, the scattering coefficients are given by the approximate expressions:

$$T_0 = -i\frac{\pi}{4}\tilde{k}_f^2 \qquad \text{and} \qquad T_1 = i\frac{\pi}{4}\tilde{k}_f^2 = -T_0 \tag{2.21}$$

and the normalized differential cross-section $\sigma_d/2R$ varies as $(R/\lambda)^3$ in the Rayleigh scattering regime:

$$\gamma_d(\theta) = \frac{\sigma_d(\theta)}{2R} = \frac{\pi}{16}\tilde{k}_f^3 \left(2\cos\theta - 1\right)^2 \tag{2.22}$$

At high frequencies ($\tilde{k}_f >> 1$), the normalized differential cross-section can be approximated by:

$$\gamma_d(\theta) = \frac{1}{4}\sin\left(\frac{\theta}{2}\right) + \frac{\sin^2\left(\tilde{k}_f\theta\right)}{4\pi\tilde{k}_f}\cot^2\left(\frac{\theta}{2}\right) \tag{2.23}$$

where the first term is a geometrical term representing the reflection by the cylinder, and the second one is a corrective term on the main lobe (Morse and Feshbach 1953).

The differential cross-section is plotted in Figure 2.3 as a function of angle θ for different values of the product $\tilde{k}_f = k_f R$. In each figure, the arrow indicates the propagation direction of the incident wave. At low frequencies, the differential scattering cross-section is small for $\theta = 0$ and maximum for $\theta = \pi$: the scattering takes place backward. Conversely, at high frequencies, the differential scattering cross-section is small for $\theta = \pi$ and maximum for $\theta = 0$: the scattering takes place forward. This result, which can be surprising, *a priori*, is explained in the following paragraph.

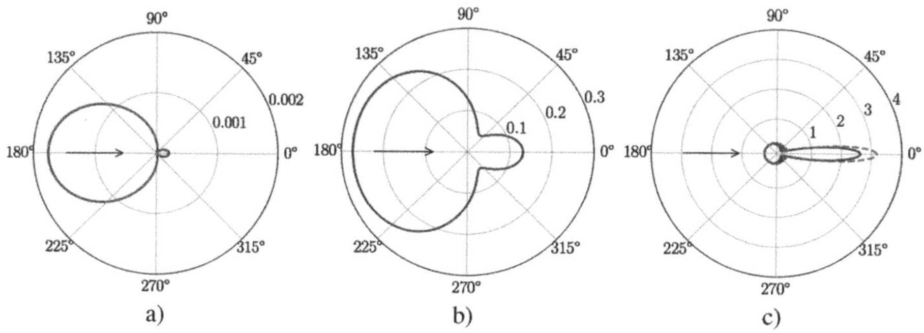

Figure 2.3. *Normalized differential scattering cross-section for a rigid cylinder of radius R and unit length, immersed in a fluid. a) $\tilde{k}_f = 0.1$, b) $\tilde{k}_f = 1$ and c) $\tilde{k}_f = 10$*

Figure 2.4 depicts the interaction, at high frequency ($\tilde{k}_f = 10$), of a longitudinal plane wave with a rigid cylinder. The map of the scattered pressure field (Figure 2.4(a)) illustrates the magnitude of the forward emission lobe shown in Figure 2.3(c). On the map of the total pressure field (Figure 2.4(b)), there is clearly a shadow area behind the cylinder. It results from the destructive interference between the incident and scattered fields, which have almost the same amplitude, and are opposite in phase. The power scattered toward the front is, therefore, equal to $2RI_{inc}$. The second part of the scattered power comes from the reflection (clearly visible in Figure 2.4(a)), which corresponds to the geometric term in expression [2.23]. The integration over θ shows that it also contributes for $2R$ to the scattering cross-section of the cylinder, whose total value is twice its geometric cross-section. This contradiction between geometrical acoustics and scattering theory only exists at high frequency. When the dimensions of the scatterer are small, with respect to the wavelength, the scattered field is a small perturbation of the incident field. This "paradox", explained by Brillouin (1949), arises from a contradiction between the intuitive association of the term "scattered" with "reflected", and the mathematical definition of the scattered field, which includes both the reflected field as well as a term that interferes with the incident field to create the shadow behind the object. Difficulties arise when we try to interpret the results of the scattering theory through intuitive concepts.

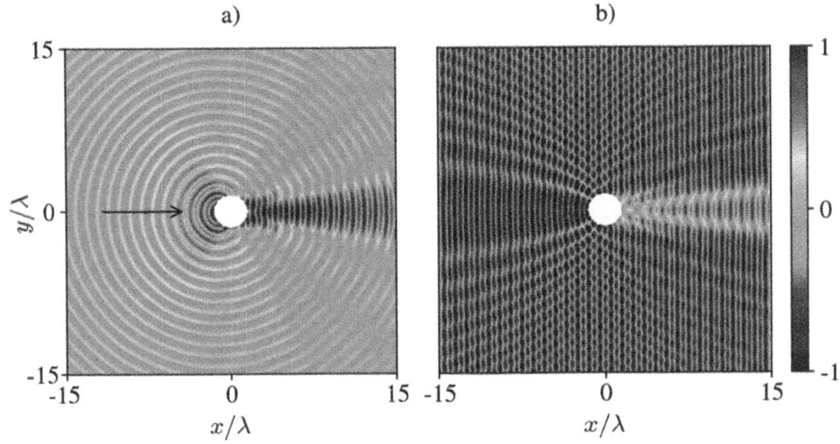

Figure 2.4. *Interaction of a plane wave with a rigid cylinder immersed in water. a) Scattered field when $\tilde{k}_f = 10$. b) Superposition of the incident and scattered fields. For a color version of this figure, see www.iste.co.uk/royer/waves2.zip*

2.1.2. *Scattering by an elastic cylinder*

The scattering by an isotropic elastic cylinder involves calculations similar to those developed for a rigid cylinder, but taking into account the longitudinal and transverse waves transmitted inside the cylinder. The cylinder material is characterized by its mass density ρ and its Lamé constants λ and μ, so that the longitudinal and transverse wave velocities are: $V_L = \sqrt{(\lambda + 2\mu)/\rho}$ and $V_T = \sqrt{\mu/\rho}$. The solution to the propagation equation is searched using the Helmholtz decomposition (equation [1.77] of Volume 1):

$$\underline{u}^p = \underline{\text{grad}}\ \Phi_L + \underline{\text{rot}}\ (\Phi_T \underline{e}_z) \qquad [2.24]$$

where the scalar potentials Φ_L and Φ_T for longitudinal and transverse waves, polarized in the xy plane, are solutions to the Helmholtz equations (see section 1.2.6.1, Volume 1):

$$\Delta\Phi_M + k_M^2 \Phi_M = 0 \qquad \text{with} \qquad \text{M} = \text{L, T} \qquad [2.25]$$

The displacement components are expressed in terms of Φ_L and Φ_T by the relations [1.126] in Volume 1:

$$u_r^p(r,\theta) = \frac{\partial \Phi_L}{\partial r} + \frac{1}{r}\frac{\partial \Phi_T}{\partial \theta} \quad \text{and} \quad u_\theta^p(r,\theta) = \frac{1}{r}\frac{\partial \Phi_L}{\partial \theta} - \frac{\partial \Phi_T}{\partial r} \qquad [2.26]$$

Only stresses σ_{rr}^p and $\sigma_{r\theta}^p$ are involved in the boundary conditions. It is therefore necessary to express them, using formulae [A1.9] of Volume 1, as functions of the scalar potentials Φ_L and Φ_T:

$$\begin{cases} \sigma_{rr}^p(r,\theta) = \dfrac{\mu}{r^2}\left[-(k_T r)^2\Phi_L - 2r\dfrac{\partial \Phi_L}{\partial r} - 2\dfrac{\partial^2 \Phi_L}{\partial \theta^2} + 2r^2\dfrac{\partial^2(\Phi_T/r)}{\partial r \partial \theta}\right] \\ \sigma_{r\theta}^p(r,\theta) = \dfrac{\mu}{r^2}\left[2r^2\dfrac{\partial^2(\Phi_L/r)}{\partial r \partial \theta} + (k_T r)^2\Phi_T + 2r\dfrac{\partial \Phi_T}{\partial r} + 2\dfrac{\partial^2 \Phi_T}{\partial \theta^2}\right] \end{cases} \qquad [2.27]$$

During the calculation, and in order to limit the order of the derivatives to one, the second derivatives of the potentials, with respect to r, have been eliminated using the Helmholtz equations.

In the fluid, the velocity potential Φ_i is defined by the development [2.6]. Since $\underline{v} = -\,\underline{\text{grad}}\;\Phi_i = -i\omega\underline{u}$, and in order to be equivalent, the displacement potentials in the solid must be divided by $i\omega$. Consequently, the scalar potentials Φ_L and Φ_T, associated with longitudinal and transverse waves, respectively, are given by Faran (1951):

$$\begin{cases} \Phi_L(r,\theta) = \dfrac{a_f}{\omega k_f}\displaystyle\sum_{n=0}^{\infty} \epsilon_n i^n A_n J_n(k_L r)\cos(n\theta) \\ \Phi_T(r,\theta) = \dfrac{a_f}{\omega k_f}\displaystyle\sum_{n=1}^{\infty} 2 i^n B_n J_n(k_T r)\sin(n\theta) \end{cases} \qquad [2.28]$$

The potential of the transverse wave, transmitted in the cylinder, is expressed as a function of $\sin(n\theta)$, while the potential of the longitudinal waves (incident, scattered, and transmitted) is expressed as a function of $\cos(n\theta)$. This difference can be explained by the derivative operation according to θ and applied to the potential Φ_T in relations [2.26] and [2.27] giving the displacement u_r^p and the stress σ_{rr}^p. From the expressions for the potentials, the normal displacement u_r^p and the stresses σ_{rr}^p and $\sigma_{r\theta}^p$ take the following forms at $r = R$:

$$\begin{cases} u_r^p(R,\theta) = \dfrac{a_f}{\omega \tilde{k}_f} \displaystyle\sum_{n=0}^{\infty} \epsilon_n i^n \left(A_n U_n^L + B_n U_n^T\right) \cos(n\theta) \\ \sigma_{rr}^p(R,\theta) = \dfrac{\mu_p a_f}{\omega \tilde{k}_f R} \displaystyle\sum_{n=0}^{\infty} \epsilon_n i^n \left(A_n \Sigma_n^L + B_n \Sigma_n^T\right) \cos(n\theta) \\ \sigma_{r\theta}^p(R,\theta) = \dfrac{\mu_p a_f}{\omega \tilde{k}_f R} \displaystyle\sum_{n=1}^{\infty} 2 i^n \left(A_n \tau_n^L + B_n \tau_n^T\right) \sin(n\theta) \end{cases} \qquad [2.29]$$

with:

$$\begin{cases} U_n^L = nJ_n(\tilde{k}_L) - \tilde{k}_L J_{n+1}(\tilde{k}_L) \\ U_n^T = nJ_n(\tilde{k}_T) \\ \Sigma_n^L = \left[2n(n-1) - \tilde{k}_T^2\right] J_n(\tilde{k}_L) + 2\tilde{k}_L J_{n+1}(\tilde{k}_L) \\ \Sigma_n^T = 2n\left[(n-1)J_n(\tilde{k}_T) - \tilde{k}_T J_{n+1}(\tilde{k}_T)\right] \\ \tau_n^L = 2n\left[(1-n)J_n(\tilde{k}_L) + \tilde{k}_L J_{n+1}(\tilde{k}_L)\right] \\ \tau_n^T = \left[\tilde{k}_T^2 + 2n(1-n)\right] J_n(\tilde{k}_T) - 2\tilde{k}_T J_{n+1}(\tilde{k}_T) \end{cases} \qquad [2.30]$$

and where $\tilde{k}_L = k_L R$ and $\tilde{k}_T = k_T R$.

The boundary conditions on the cylinder surface $r = R$ are:

$$\begin{cases} -i\omega u_r^p(R,\theta) = v_r(R,\theta) \\ \sigma_{rr}^p(R,\theta) = -p_a(R,\theta) \\ \sigma_{r\theta}^p(R,\theta) = 0 \end{cases} \qquad [2.31]$$

The substitution in the boundary conditions of the expressions of the pressure [2.8] and the radial velocity [2.9] in the fluid, as well as the displacements and stresses [2.29] in the solid, leads to a system of three equations with an infinite number of unknowns: the scattering coefficients A_n, B_n and T_n. Given the orthogonality of functions $\cos(n\theta)$ and $\sin(n\theta)$, for the harmonic $n = 0$ these equations are reduced to the following system:

$$\begin{pmatrix} U_0^L & -\tilde{k}_f H_0'(\tilde{k}_f) \\ \mu_p \Sigma_0^L & K_f(\tilde{k}_f)^2 H_0(\tilde{k}_f) \end{pmatrix} \begin{pmatrix} A_0 \\ T_0 \end{pmatrix} = \begin{pmatrix} \tilde{k}_f J_0'(\tilde{k}_f) \\ -K_f(\tilde{k}_f)^2 J_0(\tilde{k}_f) \end{pmatrix} \qquad [2.32]$$

and for each harmonic of order $n > 0$:

$$\begin{pmatrix} U_n^L & U_n^T & -\tilde{k}_f H_n'(\tilde{k}_f) \\ \mu_p \Sigma_n^L & \mu_p \Sigma_n^T & K_f(\tilde{k}_f)^2 H_n(\tilde{k}_f) \\ \tau_n^L & \tau_n^T & 0 \end{pmatrix} \begin{pmatrix} A_n \\ B_n \\ T_n \end{pmatrix} = \begin{pmatrix} \tilde{k}_f J_n'(\tilde{k}_f) \\ -K_f(\tilde{k}_f)^2 J_n(\tilde{k}_f) \\ 0 \end{pmatrix} \qquad [2.33]$$

The resolution of this system provides the scattering coefficient T_n for the waves in the fluid and, as in the previous section, the scattering cross-section. Its normalized value γ (formula [2.20]) is plotted in Figure 2.5 for three different cylinders: a rigid cylinder, a steel cylinder and a plexiglass cylinder. The scattering cross-section for the steel cylinder is relatively close to that of the rigid cylinder because its elastic moduli are very high. However, several peaks appear that correspond to resonances. There are many more resonances for the plexiglass (PMMA) cylinder, whose elastic parameters are relatively close to the bulk modulus of water $K_f = \rho_f V_f^2 \approx 2.25$ GPa. Some resonances are related to the propagation of elastic waves on the surface of the scatterer.

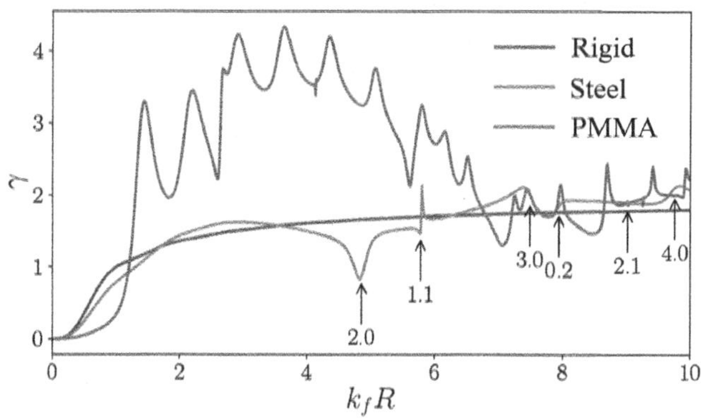

Figure 2.5. *Normalized scattering cross-section γ for a rigid cylinder, a steel cylinder and a plexiglas (PMMA) cylinder, immersed in water. For a color version of this figure, see www.iste.co.uk/royer/waves2.zip*

2.1.3. *Circumferential waves*

In the harmonic case, these waves are the solutions, finite at the origin, to the Helmholtz equations [2.25], propagating on the surface of the cylinder. The potentials Φ_L and Φ_T are given by equation [A1.17], by replacing R_n with Bessel functions of the first kind J_n, that is, for the mode of order n:

$$\begin{cases} \Phi_L(r,\theta,t) = A_n J_n(k_L r) e^{i(n\theta - \omega t)} \\ \Phi_T(r,\theta,t) = B_n J_n(k_T r) e^{i(n\theta - \omega t)} \end{cases} \qquad [2.34]$$

These exponential solutions represent waves propagating anticlockwise at the phase velocity $V = \omega R/n$. Assuming that the surface of the cylinder is traction free, the stresses σ_{rr} and $\sigma_{r\theta}$ must be canceled for $r = R$, that is, given expressions [2.29]:

$$\begin{pmatrix} \Sigma_n^L & i\Sigma_n^T \\ -i\tau_n^L & \tau_n^T \end{pmatrix} \begin{pmatrix} A_n \\ B_n \end{pmatrix} = \begin{pmatrix} 0 \\ 0 \end{pmatrix} \quad \text{and} \quad \Sigma_n^L \tau_n^T - \Sigma_n^T \tau_n^L = 0 \qquad [2.35]$$

By taking into account expressions [2.30] of coefficients Σ_n^L, Σ_n^T, τ_n^L and τ_n^T, the characteristic equation takes the form:

$$\begin{aligned} &\left[4n(n-1) - \tilde{k}_T^2\right] \tilde{k}_T^2 J_n(\tilde{k}_L) J_n(\tilde{k}_T) + 4(n^2-1) J_{n+1}(\tilde{k}_L) \tilde{k}_L \tilde{k}_T J_{n+1}(\tilde{k}_T) \\ &+2\left[\tilde{k}_T^2 + 2n(1-n^2)\right]\left[\tilde{k}_L J_{n+1}(\tilde{k}_L) J_n(\tilde{k}_T) + \tilde{k}_T J_n(\tilde{k}_L) J_{n+1}(\tilde{k}_T)\right] = 0 \end{aligned} \qquad [2.36]$$

and the ratio of the amplitudes of the potentials, for $n \neq 0$, is given by:

$$\frac{B_n}{A_n} = i\frac{\tau_n^L}{\tau_n^T} = i\frac{2n\left[(1-n)J_n(\tilde{k}_L) + \tilde{k}_L J_{n+1}(\tilde{k}_L)\right]}{\left[\tilde{k}_T^2 + 2n(1-n)\right] J_n(\tilde{k}_T) - 2\tilde{k}_T J_{n+1}(\tilde{k}_T)} \qquad [2.37]$$

The characteristic equation [2.36] is solved by denoting:

$$x = \tilde{k}_T = \frac{\omega R}{V_T} \quad \text{and} \quad y = \tilde{k}_L = \frac{\omega R}{V_L} = \kappa x \qquad [2.38]$$

For each value of n, there exists an infinite number of solutions x_m, to which correspond the angular frequencies $\omega_{n,m} = V_T x_m/R$. For each mode, the components u_r and u_θ of the mechanical displacement are deduced from the relations [2.26]. When $n = 0$ and since $\Sigma_0^T = \tau_0^L = 0$, equation [2.35] is written as $\Sigma_0^L \tau_0^T = 0$. Equation $\Sigma_0^L = 0$ is identical to the relation [4.194] in Volume 1 that gives the cut-off frequencies for the radial modes of a cylinder. Equation $\tau_0^T = 0$ is identical to the dispersion equation [4.199] of Volume 1, for the torsional modes, polarized along $\underline{e}_\theta$. The radial and tangential components of the mechanical displacement are decoupled.

Figure 2.6(a) shows the variations in the normalized frequency $\omega R/V_T$ *versus the angular wave number* kR for the steel cylinder. The integer values n of kR satisfy a phase matching condition at each revolution: $\Delta\varphi = 2\pi kR = 2n\pi$ and thus correspond to a resonance. For $n \geq 1$, the calculation of the components u_r and u_θ shows that the first root $\omega_{n,0}$ corresponds to a mechanical disturbance localized near the surface, which becomes a Rayleigh wave when $n = kR \to \infty$, that is, for a cylinder whose radius is very large compared to the wavelength. The other solutions

($m = 1, 2$, etc.) correspond to the so-called "whispering gallery modes", also discovered by Lord Rayleigh, and whose mechanical displacement is more concentrated inside the cylinder. When kR approaches unity, the frequency, and thus the phase velocity of the mode $m = 0$ cancel out (Figure 2.6(b)). The dispersion effect due to the curvature of the surface is quite large at low frequencies ($kR < 10$), that is, when the wavelength is larger than $R/2$. When kR tends to infinity, the phase velocity $V = \omega/k$ approaches the velocity V_R of the Rayleigh wave on a plane surface. It must be noted that the non-integer values of kR are involved in the transient regime to describe the non-resonant propagation of a surface disturbance around the cylinder. The number of turns is then limited by the material attenuation, the diffraction and the dispersion, which cause the wave packet to spread out.

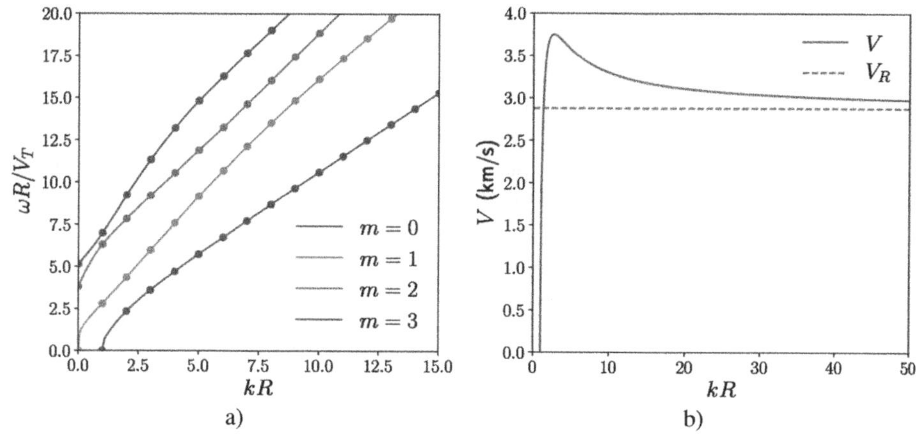

Figure 2.6. *a) Dispersion curves of the Rayleigh mode ($m = 0$) and whispering gallery modes ($m \geq 1$) on the free surface of a steel cylinder. The discrete values correspond to the resonance frequencies $\omega_{n,m}$. b) Phase velocity of the Rayleigh wave as a function of kR. For a color version of this figure, see www.iste.co.uk/royer/waves2.zip*

The position of the first circumferential modes (n, m), indicated in Figure 2.5, is in good agreement with the resonances of the scattering cross-section in the steel cylinder. Given its very small acoustic impedance, the presence of water causes little disturbance in the propagation of waves on the surface of the steel cylinder. This is no longer true for the plexiglass cylinder, and the characteristic equation of the interface waves are thus obtained by canceling the determinant of the system [2.33].

2.2. Scattering of elastic waves by a cylinder

In this section, the scatterer is an elastic cylinder embedded in an isotropic solid matrix. The displacements in the matrix and cylinder are calculated using the Helmholtz decomposition [2.24]. Let a_I (I = L, T) be the amplitude of the displacement potential for the incident plane wave (longitudinal or transverse) propagating in the x direction. Since $x = r\cos\theta$, the Jacobi and Anger development [A1.18] leads to the expression:

$$\phi^i(r,\theta) = a_I e^{ik_I x} = a_I \sum_{n=-\infty}^{\infty} i^n J_n(k_I r) e^{in\theta} \qquad \text{with} \qquad \mathrm{I = L, T} \qquad [2.39]$$

The potential associated with a scattered wave takes the form:

$$\phi^d_M(r,\theta) = a_I \sum_{n=-\infty}^{\infty} i^n T_n^{IM} H_n(k_M r) e^{in\theta} \qquad \text{with} \qquad \mathrm{M = L, T} \qquad [2.40]$$

where the scattering coefficient T_n^{IM} represents the amplitude of the cylindrical harmonic n of the scattered mode M for an incident wave of type I. The calculation of the scattering coefficients involves the boundary conditions on the cylinder surface $r = R$. This requires the calculation of displacements and stresses, given by the expressions [2.26] and [2.27], respectively.

For an incident wave of type I (L or T), the differential scattering cross-section is the sum of two terms: $\sigma_d = \sigma_L + \sigma_T$, with:

$$\sigma_L(\theta) = \frac{2}{\pi k_I}\left|\sum_{n=-\infty}^{\infty} T_n^{IL} e^{in\theta}\right|^2 \qquad \text{and} \qquad \sigma_T(\theta) = \frac{2}{\pi k_I}\left|\sum_{n=-\infty}^{\infty} T_n^{IT} e^{in\theta}\right|^2 \quad [2.41]$$

Given the orthogonality of the trigonometric functions, the scattering cross-sections normalized by the geometric cross-section of the scatterer can be written as:

$$\gamma_L = \frac{2}{\tilde{k}_I}\sum_{n=-\infty}^{\infty} |T_n^{IL}|^2 \qquad \text{and} \qquad \gamma_T = \frac{2}{\tilde{k}_I}\sum_{n=-\infty}^{\infty} |T_n^{IT}|^2 \qquad [2.42]$$

Scattering by a glass cylinder buried in an epoxy matrix is illustrated in Figure 2.7. The differential scattering cross-sections into longitudinal and transverse waves are plotted in the case of a *longitudinal incident wave* and for two values of the dimensionless wave number $\tilde{k}_L = k_L R$. The behavior of the scattered longitudinal wave is comparable with that of the acoustic wave scattered by an immersed cylinder. At low frequency ($\tilde{k}_L = 0.1$), the scattering is weak and mainly takes place backward (decreasing x). At medium frequency ($\tilde{k}_L = 1$), backward and forward (increasing x) scattering are comparable. The converted transverse waves behave similarly, but their scattering is oriented at $\pm 40°$ with respect to the propagation direction. Further, it is interesting to note that at low frequency, the differential scattering cross-section of the transverse wave is twice that of the longitudinal wave. More generally, the conversion of a longitudinal wave into a transverse wave is significant at low frequencies, as long as the cylinder is more rigid than the matrix.

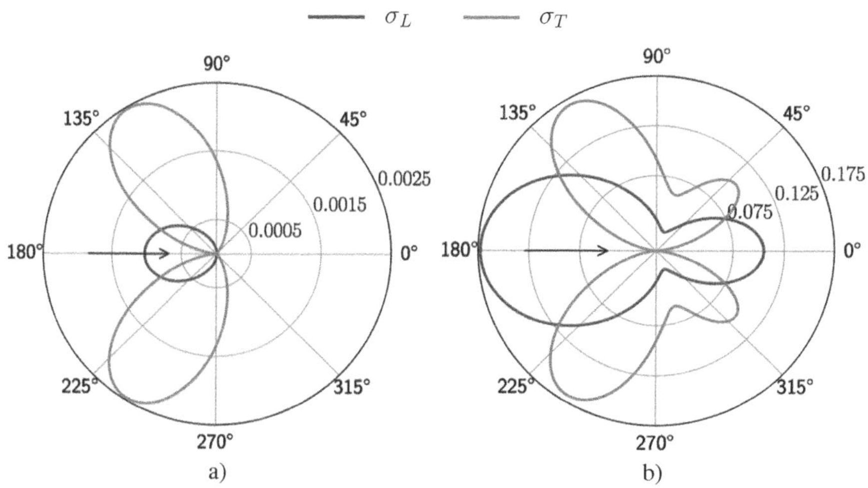

Figure 2.7. *Longitudinal incident wave. Differential scattering cross-sections for a glass cylinder in an epoxy matrix at a frequency such that a) $\tilde{k}_L = 0.1$ and b) $\tilde{k}_L = 1$. For a color version of this figure, see www.iste.co.uk/royer/waves2.zip*

The differential scattering cross-sections into longitudinal and transverse waves are plotted in Figure 2.8 in the case of a *transverse incident wave* for two values of the dimensionless wave number $\tilde{k}_T = k_T R$. The transverse wave is scattered along three main directions: backward and in two directions close to the y axis, perpendicular to the propagation direction. The converted longitudinal waves have the same behavior as the converted transverse waves presented in Figure 2.7, but with much smaller amplitudes (Liu *et al.* 2000). Wave conversion is thus smaller in the case of the transverse incident wave.

The variation of the scattering cross-sections *versus* frequency is shown in Figure 2.9 for a longitudinal and a transverse incident wave. As previously indicated, the wave conversion in the case of the longitudinal wave is significant at low frequency, that is, for $\tilde{k}_M < 1$. At higher frequencies, the scattering cross-section γ_T presents an oscillatory behavior whose period is $\Delta\tilde{k}_T \approx 4.5$ (Figure 2.9(b)).

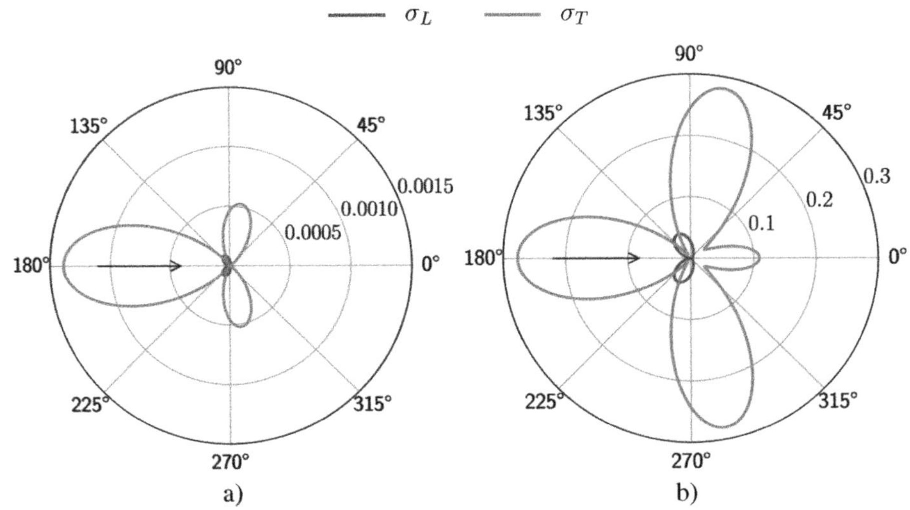

Figure 2.8. *Transverse incident wave. Differential scattering cross-section for a glass cylinder in an epoxy matrix at a frequency such that a) $\tilde{k}_T = 0.1$ and b) $\tilde{k}_T = 1$. For a color version of this figure, see www.iste.co.uk/royer/waves2.zip*

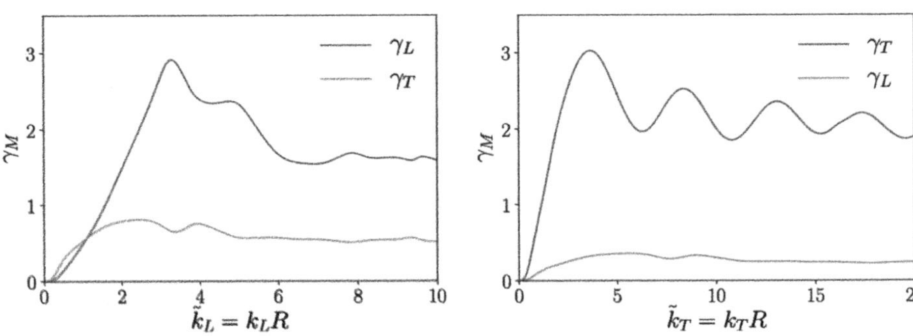

Figure 2.9. *Scattering cross-sections for a glass cylinder in an epoxy matrix. a) Longitudinal and b) transverse incident wave. For a color version of this figure, see www.iste.co.uk/royer/waves2.zip*

ORDER OF MAGNITUDE.– For a glass fiber of diameter 10 μm and a frequency $f = 5$ MHz, the reduced wave numbers are $\tilde{k}_L = 0.13$ and $\tilde{k}_T = 0.29$. In the case of an epoxy resin matrix and a longitudinal incident wave, the normalized scattering cross-sections are small: $\gamma_L = 0.004$ and $\gamma_T = 0.012$; for a transverse incident wave, the corresponding values are $\gamma_T = 0.05$ and $\gamma_L = 0.005$.

2.3. Scattering of elastic waves by a spherical particle

In the first section, we give the general expressions of the displacement components and mechanical stresses in terms of the scalar potentials. In the second and third sections, we analyze successively the interaction of a longitudinal and a transverse elastic wave with a spherical particle embedded in an isotropic matrix. The scattering by a rigid particle is first investigated and then the scattering by a gaseous or a solid elastic inclusion.

2.3.1. *General formulation of the problem*

Let R_s be the spherical coordinate system (r, θ, φ) presented in Appendix 1. The Cartesian coordinates are then defined in terms of the spherical coordinates by relations [A1.21]. The displacement field $\underline{u}$ is searched using the Helmholtz decomposition [1.186], where Φ_L, Φ_T and Φ_S are the scalar potentials, solutions to the Helmholtz equations [2.25].

Considering expressions [A1.11] and [A1.15] in Volume 1 for the operators $\underline{\underline{\text{grad}}}$ and $\underline{\text{rot}}$ in spherical coordinates, the components of the mechanical displacement are expressed as

$$\begin{cases} u_r = \dfrac{\partial \Phi_L}{\partial r} + k_T^2 r \Phi_T + r\dfrac{\partial^2 \Phi_T}{\partial r^2} + 2\dfrac{\partial \Phi_T}{\partial r} \\ u_\theta = \dfrac{1}{r}\dfrac{\partial}{\partial \theta}\left[\Phi_L + \dfrac{\partial (r\Phi_T)}{\partial r}\right] + \dfrac{k_T}{\sin\theta}\dfrac{\partial \Phi_S}{\partial \varphi} \\ u_\varphi = \dfrac{1}{r\sin\theta}\dfrac{\partial}{\partial \varphi}\left[\Phi_L + \dfrac{\partial (r\Phi_T)}{\partial r}\right] - k_T\dfrac{\partial \Phi_S}{\partial \theta} \end{cases} \quad [2.43]$$

Only the stresses σ_{rr}, $\sigma_{r\theta}$ and $\sigma_{r\varphi}$ are involved in the boundary conditions. It is therefore necessary to express these three components in terms of the scalar potentials Φ_L, Φ_T and Φ_S using formulae [A1.17] of Volume 1:

$$\begin{cases} \sigma_{rr} = -\lambda k_L^2 \Phi_L + 2\mu\dfrac{\partial^2 \Phi_L}{\partial r^2} + 2\mu\dfrac{\partial}{\partial r}\left(k_T^2 r\Phi_T + r\dfrac{\partial^2 \Phi_T}{\partial r^2} + 2\dfrac{\partial \Phi_T}{\partial r}\right) \\ \sigma_{r\theta} = \mu\left[\dfrac{\partial}{\partial \theta}\left(2\dfrac{\partial(\Phi_L/r)}{\partial r} + \mathcal{L}(\Phi_T)\right) + \dfrac{k_T r}{\sin\theta}\dfrac{\partial^2(\Phi_S/r)}{\partial r\partial\varphi}\right] \\ \sigma_{r\varphi} = \mu\left[\dfrac{1}{\sin\theta}\dfrac{\partial}{\partial\varphi}\left(2\dfrac{\partial(\Phi_L/r)}{\partial r} + \mathcal{L}(\Phi_T)\right) - k_T r\dfrac{\partial^2(\Phi_S/r)}{\partial r\partial\theta}\right] \end{cases} \quad [2.44]$$

with:

$$\mathcal{L}(\Phi_T) = 2\frac{\partial^2 \Phi_T}{\partial r^2} + \frac{2}{r}\frac{\partial \Phi_T}{\partial r} - \frac{2}{r^2}\Phi_T + k_T^2 \Phi_T \qquad [2.45]$$

In the following section, the potentials of the incident and scattered waves are developed over spherical harmonics. In Appendix 1, these harmonics are expressed using spherical Bessel functions and Legendre polynomials.

2.3.2. *Scattering of a longitudinal plane wave*

The case of a rigid sphere is treated first, in order to illustrate the decomposition of the incident and scattered fields on the spherical harmonics, as well as the use of orthogonality relations to calculate the scattering coefficients. The scattering by an elastic particle is then studied in the low-frequency domain, where monopolar and dipolar resonances appear. In the final section, the scattering cross-sections and the far-field scattering functions are calculated using approximations that are valid in the far field.

2.3.2.1. *Scattering by a rigid particle*

The incident wave is a longitudinal plane wave propagating in the z direction with the scalar potential $\Phi_L^i = (a_L/ik_L)\exp(ik_L z)$ and the displacement $\underline{u}^i = a_L \exp(ik_L z)\underline{e}_z$. As $z = r\cos\theta$, the incident wave does not depend on angle φ. Its potential may be developed over spherical harmonics of order $m = 0$, according to the formula [A1.48]:

$$\Phi_L^i(r,\theta) = \frac{a_L}{ik_L}\sum_{n=0}^{\infty} i^n(2n+1)j_n(k_L r)P_n(\cos\theta) \qquad [2.46]$$

where j_n and P_n are, respectively, the spherical Bessel functions of the first kind and the Legendre polynomials. Since the incident field is independent of angle φ, the scattered fields are also independent of φ. Moreover, the incident wave fronts, which are parallel to the xy plane (Figure 2.2(b)), impose that the displacement component u_φ is zero. Given the expressions [2.43], the potential Φ_S is necessarily zero: there is no scattering into the transverse wave polarized in the xy plane. The potentials of the scattered longitudinal and transverse waves are also developed on the spherical harmonics:

$$\begin{cases} \Phi_L^d(r,\theta) = \dfrac{a_L}{ik_L}\displaystyle\sum_{n=0}^{\infty} i^n(2n+1)T_n^{LL}h_n(k_L r)P_n(\cos\theta) \\ \Phi_T^d(r,\theta) = \dfrac{a_L}{ik_L}\displaystyle\sum_{n=1}^{\infty} i^n(2n+1)T_n^{LT}h_n(k_T r)P_n(\cos\theta) \end{cases} \qquad [2.47]$$

where T_n^{LL} and T_n^{LT} are the scattering coefficients for longitudinal and transverse waves, respectively. Note that $h_n = h_n^{(1)}$ is the spherical Hankel function of the first kind and order n corresponding to a divergent wave. For the transverse wave, the sum must necessarily start from the term $n = 1$ because there is no monopolar transverse wave (Volume 1, section 1.2.6.2). By writing $\tilde{k}_L = k_L R$ and $\tilde{k}_T = k_T R$, the expressions for the non-zero components of the displacement at $r = R$ are:

$$\begin{cases} u_r(R,\theta) = \dfrac{a_L}{i\tilde{k}_L}\displaystyle\sum_{n=0}^{\infty} i^n(2n+1)\left[U_n^{i,L} + T_n^{LL}U_n^{d,L} + T_n^{LT}U_n^{d,T}\right] P_n(\cos\theta) \\ u_\theta(R,\theta) = \dfrac{a_L}{i\tilde{k}_L}\displaystyle\sum_{n=1}^{\infty} i^n(2n+1)\left[V_n^{i,L} + T_n^{LL}V_n^{d,L} + T_n^{LT}V_n^{d,T}\right] P_n^1(\cos\theta) \end{cases} \quad [2.48]$$

with $P_n^1(\cos\theta) = \mathrm{d}P_n(\cos\theta)/\mathrm{d}\theta$ and:

$$\begin{cases} U_n^{p,L}(\tilde{k}_L) = n f_n(\tilde{k}_L) - \tilde{k}_L f_{n+1}(\tilde{k}_L) \\ U_n^{p,T}(\tilde{k}_T) = n(n+1) f_n(\tilde{k}_T) \\ V_n^{p,L}(\tilde{k}_L) = f_n(\tilde{k}_L) \\ V_n^{p,T}(\tilde{k}_T) = (1+n) f_n(\tilde{k}_T) - \tilde{k}_T f_{n+1}(\tilde{k}_T) \end{cases} \quad [2.49]$$

For the incident wave ($p = i$), the function f_n is the spherical Bessel function of the first kind j_n, while for the scattered waves ($p = d$), f_n is the spherical Hankel function of the first kind h_n. Since the particle is rigid, the displacements are zero on the surface $r = R$, leading to the system:

$$\begin{cases} \displaystyle\sum_{n=0}^{\infty} i^n(2n+1)\left[U_n^{i,L} + T_n^{LL}U_n^{d,L} + T_n^{LT}U_n^{d,T}\right] P_n(\cos\theta) = 0 \\ \displaystyle\sum_{n=1}^{\infty} i^n(2n+1)\left[V_n^{i,L} + T_n^{LL}V_n^{d,L} + T_n^{LT}V_n^{d,T}\right] P_n^1(\cos\theta) = 0 \end{cases} \quad [2.50]$$

The orthogonality relation [A1.41] for the Legendre functions allows us to decouple this system into one system for each spherical harmonic:

$$\begin{cases} U_0^{i,L} + T_0^{LL}U_0^{d,L} = 0 & \\ U_n^{i,L} + T_n^{LL}U_n^{d,L} + T_n^{LT}U_n^{d,T} = 0 & \forall n > 0 \\ V_n^{i,L} + T_n^{LL}V_n^{d,L} + T_n^{LT}V_n^{d,T} = 0 & \forall n > 0 \end{cases} \quad [2.51]$$

The scattering coefficients are then given by the relations (Gaunaurd and Überall 1976):

$$T_0^{LL} = -\frac{U_0^{i,L}}{U_0^{d,L}}, \qquad T_n^{LL} = -\frac{U_n^{i,L}V_n^{d,T} - V_n^{i,L}U_n^{d,T}}{U_n^{d,L}V_n^{d,T} - V_n^{d,L}U_n^{d,T}}$$
$$\text{and} \qquad T_n^{LT} = -\frac{U_n^{d,L}V_n^{i,L} - V_n^{d,L}U_n^{i,L}}{U_n^{d,L}V_n^{d,T} - V_n^{d,L}U_n^{d,T}} \qquad [2.52]$$

At low frequencies, the scattering coefficients of the monopolar ($n = 0$) and dipolar ($n = 1$) modes are dominant in the modal series. By performing a limited development of spherical Bessel and Hankel functions for $\tilde{k}_L << 1$, the scattering coefficients of these modes are approximately written as (Valier-Brasier *et al.* 2015):

$$T_0^{LL} = -i\frac{\tilde{k}_L^3}{3}, \quad T_1^{LL} = i\frac{\tilde{k}_L^3(\tilde{k}_T^2 + 3i\tilde{k}_T - 3)}{6\tilde{k}_T^2} \quad \text{and} \quad T_1^{LT} = -i\frac{\tilde{k}_L e^{-i\tilde{k}_T}}{2} \qquad [2.53]$$

2.3.2.2. *Scattering by an elastic particle*

In the case of a particle with mass density ρ_p and Lamé constants λ_p and μ_p, the transmission of the longitudinal and transverse waves inside the particle must necessarily be taken into account in the modeling. To do this and using the Helmholtz decomposition [1.186] with $\Phi_S = 0$, the potentials of waves transmitted in the particle are searched in the form of expansions on spherical harmonics:

$$\begin{cases} \Phi_L^p(r,\theta) = \dfrac{a_L}{ik_L}\displaystyle\sum_{n=0}^{\infty} i^n(2n+1)S_n^{LL} j_n(k_L^p r)P_n(\cos\theta) \\ \Phi_T^p(r,\theta) = \dfrac{a_L}{ik_L}\displaystyle\sum_{n=1}^{\infty} i^n(2n+1)S_n^{LT} j_n(k_T^p r)P_n(\cos\theta) \end{cases} \qquad [2.54]$$

The transmission coefficients of the waves in the particle S_n^{LL} and S_n^{LT} and the scattering coefficients in the matrix, T_n^{LL} and T_n^{LT} (see equation [2.47]), are determined by the boundary conditions.

At *low frequency*, that is, for $\tilde{k}_L << 1$, the scattering coefficients for the first two modes are given by Valier-Brasier *et al.* (2015):

$$\begin{cases} T_0^{LL} = -i\tilde{k}_L^3 \dfrac{K_p - K}{3K_p + 4\mu - \rho(\omega R)^2(1 + i\tilde{k}_L)} \\ T_1^{LL} = -i\dfrac{(1-\tilde{\rho})\tilde{k}_L^3}{3} \dfrac{\tilde{k}_T^2 + 3i\tilde{k}_T - 3}{\tilde{k}_T^2(2\tilde{\rho}+1) + 9i\tilde{k}_T - 9} \\ T_1^{LT} = i(1-\tilde{\rho})\dfrac{\tilde{k}_L\tilde{k}_T^2 e^{-i\tilde{k}_T}}{\tilde{k}_T^2(2\tilde{\rho}+1) + 9i\tilde{k}_T - 9} \end{cases} \quad [2.55]$$

The monopolar mode ($n = 0$) depends directly on the difference between the bulk moduli of the two media ($K = \lambda + \frac{2}{3}\mu$ and $K_p = \lambda_p + \frac{2}{3}\mu_p$), while the dipolar ($n = 1$) mode depends on the density ratio $\tilde{\rho} = \rho_p/\rho$. These scattering coefficients exhibit very strong resonances at low frequency, whose frequencies correspond to the cancellation of the denominators of the scattering coefficients. Moreover, it is easy to verify that the scattering coefficients [2.53] of the rigid sphere are deduced from formulae [2.55], when the parameters of the scatterer (K_p and ρ_p) are very large compared to those of the matrix (K and ρ).

2.3.2.2.1. Monopolar resonance

This resonance occurs when the angular frequency ω cancels the denominator of the scattering coefficient T_0^{LL} :

$$3K_p + 4\mu - \rho(\omega R)^2(1 + i\tilde{k}_L) = 0 \quad [2.56]$$

Assuming a rheology of Kelvin–Voigt type (equation [1.299] of Volume 1) for the shear modulus: $\mu = \mu_0 - i\omega\eta$ and by writing $(\omega R)^2\tilde{k}_L \approx \omega\omega_0^2 R^3/V_L$, we get:

$$\omega^2 + 2i\omega\left(\Gamma_{vis} + \Gamma_{rad}\right) - \omega_0^2 = 0 \quad [2.57]$$

where:

$$\Gamma_{vis} = \frac{2\eta}{\rho R^2} \quad \text{and} \quad \Gamma_{rad} = \frac{\omega_0^2 R}{2V_L} \quad [2.58]$$

with:

$$\omega_0 = \frac{1}{R}\sqrt{\frac{3K_p + 4\mu_0}{\rho}} \quad [2.59]$$

The terms Γ_{vis} and Γ_{rad} express the viscosity and radiation losses, respectively. The resonance frequency $f_0 = \omega_0/2\pi$, therefore, depends on the competition between the bulk modulus of the particle and the shear modulus of the host medium.

The complex resonance frequency, that is, the solution to equation [2.57], is given by:

$$\omega_{res} = \sqrt{\omega_0^2 - (\Gamma_{rad} + \Gamma_{vis})^2} - i\,(\Gamma_{rad} + \Gamma_{vis}) \qquad [2.60]$$

and the quality factor can be written as:

$$Q_0 = \left|\frac{\mathrm{Re}[\omega_{res}]}{2\,\mathrm{Im}[\omega_{res}]}\right| = \frac{\sqrt{\omega_0^2 - (\Gamma_{rad} + \Gamma_{vis})^2}}{2\,(\Gamma_{rad} + \Gamma_{vis})} \qquad [2.61]$$

If *the particle is gaseous*, its bulk modulus $K_p = \rho_p V_p^2$ is small. For the air, $K_p \approx 140$ kPa and the resonance is driven by the shear modulus of the solid. If the latter is too rigid, the resonance angular frequency ω_{res} is high, as well as the term Γ_{rad}. The quality factor is then small: the resonance is not efficient. Conversely, for a soft solid, that is, with a shear modulus μ_0 smaller than 100 MPa, the resonance occurs at a low frequency and the quality factor is high. However, for small bubbles (a few hundred μm), the resonance is located in the ultrasonic domain, that is, in the MHz range, where the effects of viscoelastic dissipation are significant and can no longer be represented by a Kelvin–Voigt model.

When the host medium is a viscous liquid, the static shear modulus μ_0 is zero and the modulus η corresponds to the shear viscosity (section 1.5.5, Volume 1). The resonance frequency then takes the form:

$$f_0 = \frac{\omega_0}{2\pi} = \frac{V_p}{2\pi R}\sqrt{\frac{3\rho_p}{\rho}} \qquad [2.62]$$

The resonance of a gas bubble in water, discovered by Minnaert, is very intense since its quality factor is very high:

$$Q_0 \approx \frac{\omega_0}{\Gamma_{rad} + \Gamma_{vis}} \qquad [2.63]$$

ORDER OF MAGNITUDE.– For a gas bubble in water, the typical value of the product f_0R is 3.26 m/s. The product $\tilde{k}_L = 2\pi f_0 R/V_f \approx 0.014$ is very small in front of unity: the Minnaert resonance occurs at a very low frequency.

2.3.2.2.2. Dipolar resonance

The dipolar resonance of an elastic particle occurs when the denominator of the corresponding scattering coefficients T_1^{LL} and T_1^{LT} cancels out:

$$\tilde{k}_T^2(2\tilde{\rho}+1)+9i\tilde{k}_T-9=0 \qquad [2.64]$$

Assuming that $\tilde{\rho}>5/8$, the complex root of this equation is expressed by:

$$(\tilde{k}_T)_{res}=\frac{3\sqrt{8\tilde{\rho}-5}-9i}{2(2\tilde{\rho}+1)} \qquad [2.65]$$

The resonance frequency and the quality factor [2.61] are thus given by Duranteau *et al.* (2016):

$$f_1=\frac{3V_T\sqrt{8\tilde{\rho}-5}}{4\pi R(2\tilde{\rho}+1)} \quad \text{and} \quad Q_1=\frac{1}{6}\sqrt{8\tilde{\rho}-5} \qquad [2.66]$$

2.3.2.3. *Scattering cross-section*

For large arguments, the approximate expression for the spherical Hankel function of the first kind is given by [A1.40]. The substitution of the potentials [2.47] in equations [2.43] and [2.44] allows us to express the displacements u_r^d and u_θ^d and the stresses σ_{rr}^d and $\sigma_{r\theta}^d$ of the scattered wave in the far field:

$$\begin{cases} u_r^d=a_Lf_r(\theta)\dfrac{e^{ik_Lr}}{r} \\ u_\theta^d=a_Lf_\theta(\theta)\dfrac{e^{ik_Tr}}{r} \end{cases} \quad \text{and} \quad \begin{cases} \sigma_{rr}^d=i\rho V_L\omega a_Lf_r(\theta)\dfrac{e^{ik_Lr}}{r} \\ \sigma_{r\theta}^d=i\rho V_T\omega a_Lf_\theta(\theta)\dfrac{e^{ik_Tr}}{r} \end{cases} \qquad [2.67]$$

with:

$$\begin{cases} f_r(\theta)=\dfrac{1}{ik_L}\displaystyle\sum_{n=0}^{\infty}(2n+1)T_n^{LL}P_n(\cos\theta) \\ f_\theta(\theta)=\dfrac{1}{ik_L}\displaystyle\sum_{n=1}^{\infty}(2n+1)T_n^{LT}P_n^1(\cos\theta) \end{cases} \qquad [2.68]$$

The scattered acoustic intensity, defined by:

$$\begin{aligned} I_d=\langle J_r\rangle &=-\frac{1}{2}\,\mathrm{Re}\left[\underline{\underline{\sigma}}^d\left(\underline{\dot{u}}^d\right)^\star\right]\underline{e}_r \\ &=-\frac{\omega}{2}\,\mathrm{Re}\left[i\sigma_{rr}^d\left(u_r^d\right)^\star+i\sigma_{r\theta}^d\left(u_\theta^d\right)^\star+i\sigma_{r\varphi}^d\left(u_\varphi^d\right)^\star\right] \end{aligned} \qquad [2.69]$$

is written as follows in the far field, with $u^d_\varphi = 0$ and $\kappa = V_T/V_L$:

$$I_d(r,\theta) = \frac{I_{inc}}{r^2}\left[|f_r(\theta)|^2 + \kappa\,|f_\theta(\theta)|^2\right] \quad \text{with} \quad I_{inc} = \frac{1}{2}\rho V_L \omega^2 |a_L|^2 \qquad [2.70]$$

According to equation [2.2], the scattered power is the average intensity over a sphere of arbitrary large radius r:

$$\begin{aligned} P_d &= 2\pi \int_0^\pi I_d(r,\theta) r^2 \sin\theta \,\mathrm{d}\theta \\ &= 2\pi I_{inc} \int_0^\pi \left[|f_r(\theta)|^2 + \kappa\,|f_\theta(\theta)|^2\right] \sin\theta\,\mathrm{d}\theta \end{aligned} \qquad [2.71]$$

The scattering cross-section $\sigma = P_d/I_{inc}$ is the sum of two terms: $\sigma = \sigma_L + \sigma_T$. Using the orthogonality relation [A1.41], we get:

$$\begin{cases} \sigma_L = \dfrac{4\pi}{k_L^2}\displaystyle\sum_{n=0}^{\infty}(2n+1)\,|T_n^{LL}|^2 \\ \sigma_T = \dfrac{4\pi}{k_L k_T}\displaystyle\sum_{n=1}^{\infty} n(n+1)(2n+1)\,|T_n^{LT}|^2 \end{cases} \qquad [2.72]$$

or again, by normalizing by the geometric cross-section of the scatterer $S = \pi R^2$:

$$\begin{cases} \gamma_L = \dfrac{4}{\tilde{k}_L^2}\displaystyle\sum_{n=0}^{\infty}(2n+1)\,|T_n^{LL}|^2 \\ \gamma_T = \dfrac{4}{\tilde{k}_L \tilde{k}_T}\displaystyle\sum_{n=1}^{\infty} n(n+1)(2n+1)\,|T_n^{LT}|^2 \end{cases} \qquad [2.73]$$

The scattering of a longitudinal plane wave by an elastic sphere buried in an epoxy resin is shown in Figure 2.10. The normalized cross-sections γ_L and γ_T have similar values in the case of the glass sphere and exhibit a maxima at slightly different frequencies. The scattering cross-section of a transverse wave for a lead sphere reaches a maximum value for the dimensionless wave number $\tilde{k}_L \approx 0.3$, which corresponds to the dipolar resonance frequency f_1 of the translational motion of the sphere in epoxy (Figure 2.10(b)). The scattering cross-section of the longitudinal wave is much smaller than that of the transverse wave, signifying that the energy carried by the incident wave is preferentially scattered into the transverse wave rather than into the longitudinal wave. It should be noted that each of these coefficients approaches unity at high frequency and, therefore, the total cross-section has a limit equal to $2\pi R^2$, as predicted.

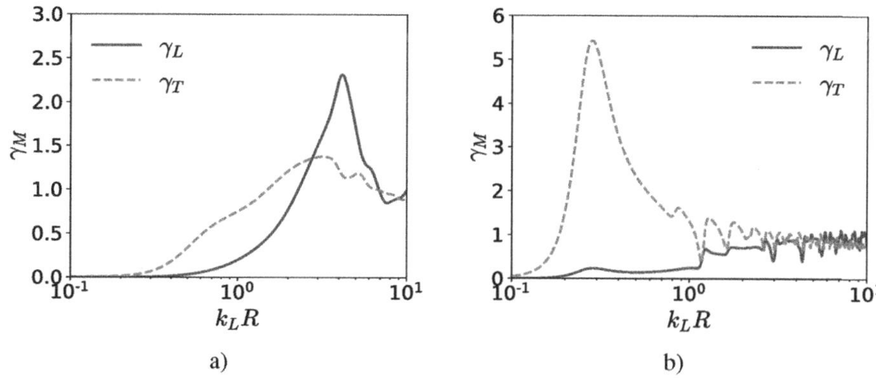

Figure 2.10. *Longitudinal incident wave. Normalized scattering cross-section for a) a glass bead and b) a lead bead, buried in an epoxy resin*

The moduli of the far-field scattering functions, f_r and f_θ, are plotted in the case of a glass bead for $\tilde{k}_L = 4.3$ (Figure 2.11(a)) and a lead bead for $\tilde{k}_L = 0.3$ (Figure 2.11(b)). For the glass bead, the far-field scattering function f_r, which depends only on the scattering coefficients T_n^{LL}, is highly directional: the longitudinal scattering takes place in the forward direction. In contrast, the scattering of the transverse wave, represented by the function f_θ, is less directional and much smaller. For the lead bead, the two far-field scattering functions present dipolar behavior (Figure 2.11(b)). This result is quite logical, given that these functions are calculated at the translational resonance frequency f_1 of the particle. On the other hand, the waves are scattered in different directions: scattering takes place preferentially in the propagation direction for the longitudinal wave, while it occurs perpendicularly for the transverse wave. This behavior results from the translational motion of the solid bead, which follows the polarization of the incident wave.

2.3.3. *Scattering of a transverse plane wave*

The analysis of the scattering of a transverse wave by a spherical particle is more complicated than the scattering of an incident longitudinal wave, because of the polarization perpendicular to the propagation direction. The calculation of a transverse plane wave in the spherical coordinate basis is first carried out using Helmholtz decomposition, which requires two scalar potentials. The scattering by a rigid sphere is then examined; the decomposition of the incident and scattered fields on the spherical harmonics shows that one of the transverse waves is coupled with the converted longitudinal wave, while the other is decoupled. In the last section, the scattering cross-sections and far-field scattering functions are calculated and the resonances arising from the propagation of the Stoneley wave on the surface of a lead sphere are highlighted.

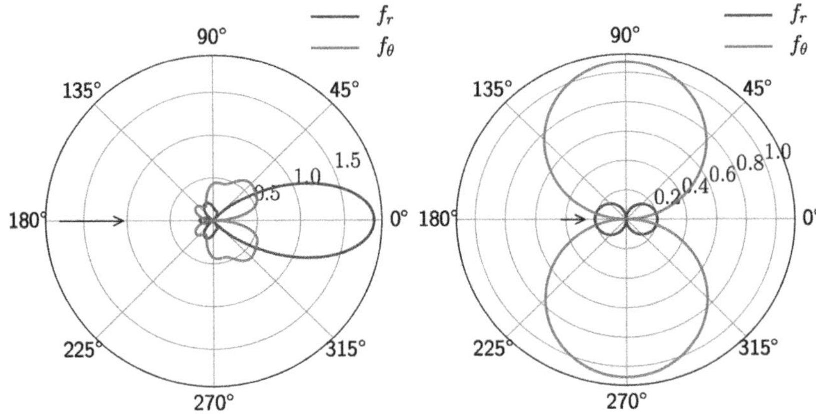

a) b)

Figure 2.11. *Far field scattering functions for a bead made of a) glass ($\tilde{k}_L = 4.3$) and b) lead ($\tilde{k}_L = 0.3$) buried in an epoxy resin. The circles of equal amplitude are graduated in mm. For a color version of this figure, see www.iste.co.uk/royer/waves2.zip*

2.3.3.1. *Transverse incident plane wave*

We are interested in the scattering by a sphere of a plane transverse wave propagating in the z direction and polarized along x. A vector potential parallel to y:

$$\underline{\psi}(z) = e^{ik_T z}\underline{e}_y \qquad [2.74]$$

imposes a particle displacement $\underline{u}$ oriented in the x direction:

$$\underline{u} = \underline{\text{rot}}\ \underline{\psi} = -ik_T e^{ik_T z}\underline{e}_x \qquad [2.75]$$

Given the expression for the vector $\underline{e}_y$:

$$\underline{e}_y = \sin\theta\sin\varphi\underline{e}_r + \cos\theta\sin\varphi\underline{e}_\theta + \cos\varphi\underline{e}_\varphi \qquad [2.76]$$

the components of the vector $\underline{\psi}$ are defined in the spherical coordinate system through the relations:

$$\begin{cases} \psi_r(r,\theta,\varphi) = e^{ik_T r\cos\theta}\sin\theta\sin\varphi \\ \psi_\theta(r,\theta,\varphi) = e^{ik_T r\cos\theta}\cos\theta\sin\varphi \\ \psi_\varphi(r,\theta,\varphi) = e^{ik_T r\cos\theta}\cos\varphi \end{cases} \qquad [2.77]$$

The vector potential $\underline{\psi}$ is expressed in terms of two scalar potentials ϕ_T and ϕ_S by the relation [1.186]:

$$\underline{\psi} = \underline{\mathrm{rot}}\,(r\phi_T\underline{e}_r) + \underline{\mathrm{rot}}\left[\underline{\mathrm{rot}}\left(\frac{r}{k_T}\phi_S\underline{e}_r\right)\right] \qquad [2.78]$$

that is:

$$\begin{cases} \psi_r = -\dfrac{1}{k_T r\sin\theta}\left[\dfrac{\partial}{\partial\theta}\left(\sin\theta\dfrac{\partial\phi_S}{\partial\theta}\right) + \dfrac{1}{\sin\theta}\dfrac{\partial^2\phi_S}{\partial\varphi^2}\right] \\ \psi_\theta = \dfrac{1}{\sin\theta}\dfrac{\partial\phi_T}{\partial\varphi} + \dfrac{1}{k_T r}\dfrac{\partial}{\partial r}\left(\dfrac{\partial r\phi_S}{\partial\theta}\right) \\ \psi_\varphi = -\dfrac{\partial\phi_T}{\partial\theta} + \dfrac{1}{k_T r\sin\theta}\dfrac{\partial}{\partial r}\left(r\dfrac{\partial\phi_S}{\partial\varphi}\right) \end{cases} \qquad [2.79]$$

By analogy with the development [A1.48] of a plane wave, the scalar potentials ϕ_T and ϕ_S are expanded on the spherical harmonics $P_n^1(\cos\theta)\cos\varphi$ or $P_n^1(\cos\theta)\sin\varphi$, whose dependence on φ is similar to the dependence of ψ_θ and ψ_φ, given the derivatives:

$$\begin{cases} \phi_T(r,\theta,\varphi) = \displaystyle\sum_{n=1}^{\infty} A_n i^n j_n(k_T r)P_n^1(\cos\theta)\cos\varphi \\ \phi_S(r,\theta,\varphi) = \displaystyle\sum_{n=1}^{\infty} B_n i^n j_n(k_T r)P_n^1(\cos\theta)\sin\varphi \end{cases} \qquad [2.80]$$

where the amplitudes A_n and B_n are to be identified. The spherical Bessel functions of the second kind are not taken into account in these two developments because they are not defined at $r = 0$. Knowing that the function $P_n^1(\cos\theta)$ is a solution to the associated Legendre equation [A1.29]:

$$\frac{\mathrm{d}}{\mathrm{d}\theta}\left(\sin\theta\frac{\mathrm{d}P_n^1(\cos\theta)}{\mathrm{d}\theta}\right) + n(n+1)P_n^1(\cos\theta)\sin\theta = \frac{P_n^1(\cos\theta)}{\sin\theta} \qquad [2.81]$$

the radial component of the potential $\underline{\psi}$ takes the form:

$$\psi_r = \sum_{n=1}^{\infty} n(n+1)B_n i^n \frac{j_n(k_T r)}{k_T r} P_n^1(\cos\theta)\sin\varphi \qquad [2.82]$$

Given the spherical harmonic expansion [A1.48] of the plane wave and since $P_0(\cos\theta) = 1$ and $\mathrm{d}P_n(\cos\theta)/\mathrm{d}\theta = P_n^1(\cos\theta)$, we get, after derivation with respect to θ:

$$\sin\theta e^{ik_T r\cos\theta} = \frac{i}{k_T r}\sum_{n=1}^{\infty}(2n+1)i^n j_n(k_T r)P_n^1(\cos\theta) \qquad [2.83]$$

Comparing expressions [2.77] and [2.82] for the component ψ_r leads to:

$$B_n = i\frac{(2n+1)}{n(n+1)} \qquad [2.84]$$

Given the relation [2.77] and substituting the potentials [2.80] into the expressions [2.79] of ψ_θ and ψ_φ leads to:

$$\cos\theta e^{ik_T r\cos\theta} = \sum_{n=1}^{\infty} i^n\left[\frac{B_n}{k_T r}\frac{\mathrm{d}\left[rj_n(k_T r)\right]}{\mathrm{d}r}\frac{\mathrm{d}P_n^1(\cos\theta)}{\mathrm{d}\theta} - A_n j_n(k_T r)\frac{P_n^1(\cos\theta)}{\sin\theta}\right] \qquad [2.85]$$

and:

$$e^{ik_T r\cos\theta} = \sum_{n=1}^{\infty} i^n\left[\frac{B_n}{k_T r}\frac{\mathrm{d}\left[rj_n(k_T r)\right]}{\mathrm{d}r}\frac{P_n^1(\cos\theta)}{\sin\theta} - A_n j_n(k_T r)\frac{\mathrm{d}P_n^1(\cos\theta)}{\mathrm{d}\theta}\right] \qquad [2.86]$$

The first term on the right-hand side of these two equations is eliminated by multiplying the first equation by $P_n^1(\cos\theta)$ and the second one by $\left[\mathrm{d}P_n^1(\cos\theta)/\mathrm{d}\theta\right]\sin\theta$, and then by integrating over θ, from 0 to π. Indeed, it appears the integral:

$$\int_0^\pi P_n^1(\cos\theta)\frac{\mathrm{d}P_n^1(\cos\theta)}{\mathrm{d}\theta}\,\mathrm{d}\theta = \frac{1}{2}\left[\left(P_n^1(\cos\theta)\right)^2\right]_0^\pi = 0 \qquad [2.87]$$

which is zero, since the Legendre functions are either even or odd. Adding each side of the transformed equations [2.85] and [2.86] leads to:

$$\begin{aligned}&\int_0^\pi e^{ik_T r\cos\theta}\left[\cos\theta P_n^1(\cos\theta) + \frac{\mathrm{d}P_n^1(\cos\theta)}{\mathrm{d}\theta}\sin\theta\right]\mathrm{d}\theta\\ &= -\sum_{n=1}^{\infty} A_n i^n j_n(k_T r)\int_0^\pi\left[\left(\frac{P_n^1(\cos\theta)}{\sin\theta}\right)^2 + \left(\frac{\mathrm{d}P_n^1(\cos\theta)}{\mathrm{d}\theta}\right)^2\right]\sin\theta\,\mathrm{d}\theta\end{aligned} \qquad [2.88]$$

Using the spherical harmonic expansion [A1.48] of the plane wave and noting that the term between square brackets is equal to $\mathrm{d}[P_n^1(\cos\theta)\sin\theta]/\,\mathrm{d}\theta$, the integral of the left-hand side is calculated by parts:

$$\begin{aligned}&\int_0^\pi P_n(\cos\theta)\left[\cos\theta P_n^1(\cos\theta)+\frac{\mathrm{d}P_n^1(\cos\theta)}{\mathrm{d}\theta}\sin\theta\right]\mathrm{d}\theta\\&=-\int_0^\pi\left[P_n^1(\cos\theta)\right]^2\sin\theta\,\mathrm{d}\theta=-\frac{2n(n+1)}{2n+1}\end{aligned}\qquad[2.89]$$

considering the normalization [A1.41] of the Legendre functions $P_n^1(\cos\theta)$. Given relation [A1.43], the equality of each term of the sum in equation [2.88] leads to:

$$A_n=\frac{2n+1}{n(n+1)}\qquad\text{and}\qquad B_n=iA_n\qquad(n\geq 1)\qquad[2.90]$$

2.3.3.2. *Scattering coefficients for a rigid sphere*

The incident wave is a transverse plane wave propagating in the z direction with the vector potential $\underline{\Psi}_i=(ia_T/k_T)\exp(ik_Tz)\underline{e}_y$ and mechanical displacement $\underline{u}_i=a_T\exp(ik_Tz)\underline{e}_x$. Three scattered waves are considered in the host medium; they are represented by the scalar potentials:

$$\begin{cases}\Phi_M^d(r,\theta)=\dfrac{ia_T}{k_T}\displaystyle\sum_{n=1}^{\infty}i^nA_nT_n^{TM}h_n(k_Mr)P_n^1(\cos\theta)\cos\varphi\quad\text{with}\quad \mathrm{M=L,T}\\\Phi_S^d(r,\theta)=-\dfrac{a_T}{k_T}\displaystyle\sum_{n=1}^{\infty}i^nA_nT_n^{SS}h_n(k_Tr)P_n^1(\cos\theta)\sin\varphi\end{cases}\qquad[2.91]$$

where the scattering coefficients T_n^{TL}, T_n^{TT} and T_n^{SS} are to be determined. The components of the mechanical displacement on the sphere surface are expressed by:

$$\begin{cases}u_r=\dfrac{ia_T}{\tilde{k}_T}\displaystyle\sum_{n=1}^{\infty}i^nA_n\left(U_n^{i,T}+T_n^{TL}U_n^{d,L}+T_n^{TT}U_n^{d,T}\right)P_n^1(\cos\theta)\cos\varphi\\u_\theta=\dfrac{ia_T}{\tilde{k}_T}\displaystyle\sum_{n=1}^{\infty}i^nA_n\left[\left(V_n^{i,T}+T_n^{TL}V_n^{d,L}+T_n^{TT}V_n^{d,T}\right)\dfrac{\mathrm{d}P_n^1(\cos\theta)}{\mathrm{d}\theta}\right.\\\qquad\left.+\left(W_n^{i,S}+T_n^{SS}W_n^{d,S}\right)\dfrac{P_n^1(\cos\theta)}{\sin\theta}\right]\cos\varphi\\u_\varphi=-\dfrac{ia_T}{\tilde{k}_T}\displaystyle\sum_{n=1}^{\infty}i^nA_n\left[\left(V_n^{i,T}+T_n^{TL}V_n^{d,L}+T_n^{TT}V_n^{d,T}\right)\dfrac{P_n^1(\cos\theta)}{\sin\theta}\right.\\\qquad\left.+\left(W_n^{i,S}+T_n^{SS}W_n^{d,S}\right)\dfrac{\mathrm{d}P_n^1(\cos\theta)}{\mathrm{d}\theta}\right]\sin\varphi\end{cases}\qquad[2.92]$$

where the coefficients $U_n^{p,M}$ and $V_n^{p,M}$, with $p = i, d$ and M = L, T, are given by expressions [2.49] and the coefficients $W_n^{p,S}$ by:

$$W_n^{(p,S)} = i\tilde{k}_T z_n(\tilde{k}_T) \tag{2.93}$$

The use of the orthogonality relations [A1.41], [A1.43] and [A1.44] allows us to decouple the system resulting from the cancellation of the three displacement components, given by relations [2.92], into one system for each harmonic n:

$$\begin{cases} U_n^{i,T} + T_n^{TL} U_n^{d,L} + T_n^{TT} U_n^{d,T} = 0 \\ V_n^{i,T} + T_n^{TL} V_n^{d,L} + T_n^{TT} V_n^{d,T} = 0 \\ W_n^{i,S} + T_n^{SS} W_n^{d,S} = 0 \end{cases} \tag{2.94}$$

Since the third equation is independent of the first two, the transverse wave S is decoupled from the L and T waves. The scattering coefficients T_n^{TT}, T_n^{TL} and T_n^{SS} are expressed by:

$$T_n^{TL} = -\frac{U_n^{i,T} V_n^{d,T} - V_n^{i,T} U_n^{d,T}}{U_n^{d,L} V_n^{d,T} - V_n^{d,L} U_n^{d,T}}, \qquad T_n^{TT} = -\frac{U_n^{d,L} V_n^{i,T} - V_n^{d,L} U_n^{i,T}}{U_n^{d,L} V_n^{d,T} - V_n^{d,L} U_n^{d,T}} \tag{2.95}$$

and $T_n^{SS} = -\dfrac{W_n^{i,S}}{W_n^{d,S}}$

2.3.3.3. *Scattering cross-section*

In the far field, considering expressions [2.43] and [2.44], the components of the displacement and stresses for the scattered field are expressed as:

$$\begin{cases} u_r^d = a_T f_r(\theta,\varphi)\dfrac{e^{ik_L r}}{r} \\ u_\theta^d = a_T f_\theta(\theta,\varphi)\dfrac{e^{ik_T r}}{r} \\ u_\varphi^d = a_T f_\varphi(\theta,\varphi)\dfrac{e^{ik_T r}}{r} \end{cases} \quad \text{and} \quad \begin{cases} \sigma_{rr}^d = i\rho V_L \omega a_T f_r(\theta,\varphi)\dfrac{e^{ik_L r}}{r} \\ \sigma_{r\theta}^d = i\rho V_T \omega a_T f_\theta(\theta,\varphi)\dfrac{e^{ik_T r}}{r} \\ \sigma_{r\varphi}^d = i\rho V_T \omega a_T f_\varphi(\theta,\varphi)\dfrac{e^{ik_T r}}{r} \end{cases} \tag{2.96}$$

with:

$$\begin{cases} f_r(\theta,\varphi) = -\dfrac{1}{ik_T}\displaystyle\sum_{n=1}^{\infty} A_n T_n^{TL} P_n^1(\cos\theta)\cos\varphi \\ f_\theta(\theta,\varphi) = -\dfrac{1}{ik_T}\displaystyle\sum_{n=1}^{\infty} A_n \left[T_n^{TT}\dfrac{\mathrm{d}P_n^1(\cos\theta)}{\mathrm{d}\theta} + T_n^{SS}\dfrac{P_n^1(\cos\theta)}{\sin\theta}\right]\cos\varphi \\ f_\varphi(\theta,\varphi) = \dfrac{1}{ik_T}\displaystyle\sum_{n=1}^{\infty} A_n \left[T_n^{TT}\dfrac{P_n^1(\cos\theta)}{\sin\theta} + T_n^{SS}\dfrac{\mathrm{d}P_n^1(\cos\theta)}{\mathrm{d}\theta}\right]\sin\varphi \end{cases} \quad [2.97]$$

Expression [2.69] for the scattered intensity takes the form:

$$I_d(r,\theta,\varphi) = \frac{I_{inc}}{r^2}\left[\kappa^{-1}|f_r(\theta,\varphi)|^2 + |f_\theta(\theta,\varphi)|^2 + |f_\varphi(\theta,\varphi)|^2\right] \quad [2.98]$$

with:

$$I_{inc} = \frac{1}{2}\rho V_T \omega^2 |a_T|^2 \quad [2.99]$$

The scattered power is the average of $I_d(r,\theta,\varphi)$ over a sphere of arbitrary large radius r. Integrating over angle φ from 0 to 2π amounts to multiplying each of the three terms by π:

$$P_d = P_L^d + P_T^d + P_S^d \quad [2.100]$$

with:

$$\begin{cases} P_L^d = \dfrac{\pi I_{inc}}{k_L k_T}\displaystyle\int_0^\pi \left|\sum_{n=1}^{\infty} A_n T_n^{TL} P_n^1(\cos\theta)\right|^2 \sin\theta\,\mathrm{d}\theta \\ P_T^d = \dfrac{\pi I_{inc}}{k_T^2}\displaystyle\int_0^\pi \left|\sum_{n=1}^{\infty} A_n T_n^{TT}\left[\dfrac{\mathrm{d}P_n^1(\cos\theta)}{\mathrm{d}\theta} + \dfrac{P_n^1(\cos\theta)}{\sin\theta}\right]\right|^2 \sin\theta\,\mathrm{d}\theta \\ P_S^d = \dfrac{\pi I_{inc}}{k_T^2}\displaystyle\int_0^\pi \left|\sum_{n=1}^{\infty} A_n T_n^{SS}\left[\dfrac{P_n^1(\cos\theta)}{\sin\theta} + \dfrac{\mathrm{d}P_n^1(\cos\theta)}{\mathrm{d}\theta}\right]\right|^2 \sin\theta\,\mathrm{d}\theta \end{cases} \quad [2.101]$$

Given the orthogonality relations of the Legendre functions [A1.41] to [A1.44] and the normalization by πR^2 and by the intensity I_{inc} of the incident plane wave, the scattering cross-sections $\gamma_M = P^d_M/(\pi R^2 I_{inc})$ are expressed by Einspruch *et al.* (1960):

$$\begin{cases} \gamma_L = \dfrac{2}{\tilde{k}_L \tilde{k}_T} \displaystyle\sum_{n=1}^{\infty} \dfrac{2n+1}{n(n+1)} \left|T_n^{TL}\right|^2 \\ \gamma_T = \dfrac{2}{\tilde{k}_T^2} \displaystyle\sum_{n=1}^{\infty} (2n+1) \left|T_n^{TT}\right|^2 \\ \gamma_S = \dfrac{2}{\tilde{k}_T^2} \displaystyle\sum_{n=1}^{\infty} (2n+1) \left|T_n^{SS}\right|^2 \end{cases} \qquad [2.102]$$

Figure 2.12 illustrates the scattering of a transverse plane wave by an elastic sphere buried in an epoxy resin. For the steel sphere (Figure 2.12(a)), the scattering cross-sections for the longitudinal and transverse T waves exhibit a maximum for the dimensionless wave number $\tilde{k}_T \approx 0.83$, corresponding to the dipolar resonance frequency f_1 of the translational motion of the sphere in epoxy. The scattering cross-section of the transverse S wave has a maximum for the dimensionless wave number $\tilde{k}_T \approx 1.32$. This peak is associated with the rotational dipolar resonance of the particle. The same resonances are visible for the lead sphere (Figure 2.12(b)), but with larger amplitudes, since lead is denser than steel and therefore the quality factor [2.66] associated with its resonance is larger. In addition, it appears as a series of peaks, almost regularly spaced apart by a value $\Delta\tilde{k}_T = 0.71$ for higher orders. These resonances are due to the existence of a Stoneley wave propagating at the interface between the lead and the epoxy with the velocity $V_{Sto} = 651$ m/s when this interface is plane. This value, that is, the solution to the characteristic equation $d_{Sto} = 0$ ([3.128], Volume 1), is the limit of the velocity of the circumferential Stoneley wave, when $k_T R >> 1$.

Given the transverse velocity in epoxy $V_T = 953$ m/s, the gap between two peaks, corresponding to $\Delta(k_{Sto}R) = 1$ with $k_{Sto} = \omega/V_{Sto}$, related to a complete turn around the sphere, tends to:

$$\Delta(k_T R) = \frac{V_{Sto}}{V_T}\Delta(k_{Sto}R) = \frac{V_{Sto}}{V_T} = 0,68 \quad \text{when} \quad \tilde{k}_T = k_T R \to \infty \qquad [2.103]$$

The characteristic equation $d_{Sto} = 0$ has no solution for the steel-epoxy couple. There is no Stoneley wave in this case, which explains the absence of resonances due to circumferential waves in Figure 2.12(a).

2.4. Scattering by a set of particles

A wave propagating in a heterogeneous medium is scattered by all obstacles it encounters. If the number of scatterers present on its path is small, the interactions between the fields scattered by each obstacle may not be taken into account. This *simple scattering* regime is used in certain imaging techniques based on the measurement of the paths of direct waves. If the wave encounters several obstacles, the multiple interactions between all the scattered fields lead to the *multiple scattering* regime. In a dilute medium, that is, with a low concentration of heterogeneities, coherent waves can be observed. These waves, which are the part of the field that is resistant to averaging over the disorder, are attenuated and their velocity is close to that of the host medium; they propagate as in a homogeneous medium, both dissipative and dispersive, called an *effective medium*.

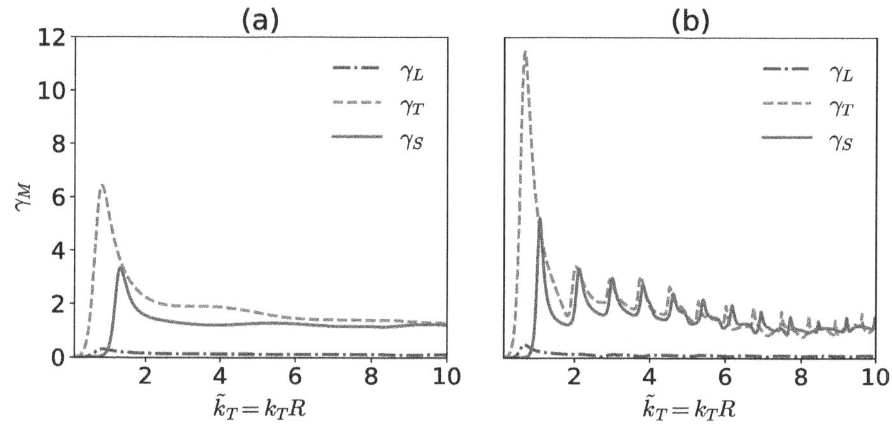

Figure 2.12. *Transverse incident wave. Scattering cross-section for a) a steel bead and b) a lead bead, embedded in an epoxy resin. For a color version of this figure, see www.iste.co.uk/royer/waves2.zip*

2.4.1. *Scalar theory of multiple scattering*

Let us study the propagation, in a heterogeneous medium, of a wave represented by the scalar potential Φ_{inc}, for example, a longitudinal wave in a perfect fluid. By hypothesis, the distribution is monodisperse in size and shape: the obstacles are all identical and randomly distributed in the volume V. The interaction of the incident wave with the N obstacles positioned at $\underline{x}_m$ (with $m = 1,..,N$) leads to the generation of N scattered waves of potentials $\Phi_d^{(m)}$. At point $\underline{x}$, the potential representing the total field resulting from the propagation of all the waves may thus be written in the form:

$$\Phi(\underline{x}) = \Phi_{inc}(\underline{x}) + \sum_{m=1}^{N} \Phi_d^{(m)}(\underline{x}, \underline{x}_m) \qquad [2.104]$$

This potential takes into account the primary interactions between the incident wave and the obstacles but also all the secondary interactions between the fields scattered by each obstacle (Figure 2.13).

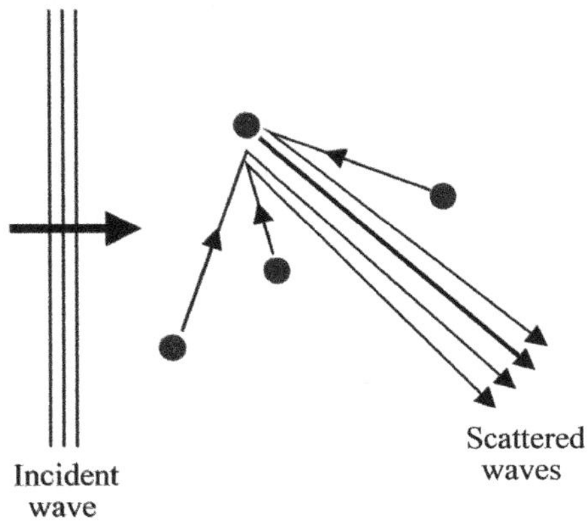

Figure 2.13. *The field scattered by an obstacle at a point in space is the sum of the simple scattering (thick lines) and the waves that have undergone multiple scattering*

The incident field on the pth scatterer can be written as the sum of the incident field Φ_{inc} and the fields scattered by all the other obstacles:

$$\Phi_i^{(p)}(\underline{x}, \underline{x}_p) = \Phi_{inc}(\underline{x}) + \sum_{\substack{m=1 \\ m \neq p}}^{N} \Phi_d^{(m)}(\underline{x}, \underline{x}_m) \qquad [2.105]$$

The key for solving this problem is to determine the transition operator $T(\underline{x}_p)$ that connects the incident field arriving at the pth scatterer to the field scattered by this obstacle:

$$\Phi_d^{(p)}(\underline{x}, \underline{x}_p) = T(\underline{x}_p)\Phi_i^{(p)}(\underline{x}, \underline{x}_p) \qquad [2.106]$$

Using this relation, without knowing, at this stage, the expression for $T(\underline{x}_p)$, equations [2.104] and [2.105] are written in the form:

$$\Phi(\underline{x}) = \Phi_{inc}(\underline{x}) + \sum_{m=1}^{N} T(\underline{x}_m)\Phi_i^{(m)}(\underline{x}, \underline{x}_m) \qquad [2.107]$$

and:

$$\Phi_i^{(p)}(\underline{x}, \underline{x}_p) = \Phi_{inc}(\underline{x}) + \sum_{\substack{m=1 \\ m \neq p}}^{N} T(\underline{x}_m)\Phi_i^{(m)}(\underline{x}, \underline{x}_m) \qquad [2.108]$$

Several strategies can be considered for solving the system of equations [2.107] and [2.108]. A first method consists of solving the problem by using the addition theorem to write the incident fields on each scatterer in its own reference frame (Rohfritsch *et al.* 2019). This procedure requires the inversion of large linear systems, which can generally only be done numerically. An example of the result is shown in Figure 2.14(a), which illustrates the field scattered by a plane wave propagating in the direction of increasing x and interacting with a random distribution of 400 steel cylinders immersed in water, between the two planes represented by the dashed vertical lines.

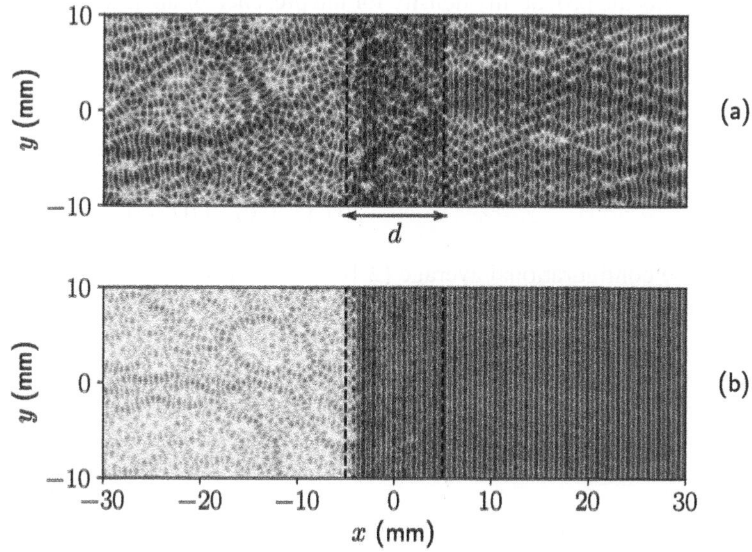

Figure 2.14. *Field scattered by a set of steel cylinders immersed in water. a) Field for a random draw and b) field averaged over 20 draws. The cylinders are arranged between the two vertical planes depicted by the dashed lines. For a color version of this figure, see www.iste.co.uk/royer/waves2.zip*

The field averaged over 20 different distributions of the positions of the cylinders is presented in Figure 2.14(b). This shows the creation of a coherent wave front, inside the layer, propagating in the direction of increasing x. This result illustrates another way to approach this problem, which consists of searching for the effective properties of coherent waves. For this purpose, the fields are averaged over all possible positions of the scatterers. In this perspective, let us consider a random distribution of N scatterers, characterized by a probability density function $p(\underline{x}_1, \underline{x}_2, ..., \underline{x}_N)$. The configurational average is defined by the relation:

$$\langle f(\underline{x})\rangle_N = \int_V f(\underline{x}\,|\underline{x}_1, ..., \underline{x}_N)p(\underline{x}_1, ..., \underline{x}_N)\,\mathrm{d}\underline{x}_1 ... \,\mathrm{d}\underline{x}_N \quad [2.109]$$

Physically, the average field $\langle f(\underline{x})\rangle$ is the statistical mean of the field $f(\underline{x})$ over all possible positions of the scatterers in the volume V. The probability density function $p(\underline{x}_1, ..., \underline{x}_N)$ is put in the form:

$$p\,(\underline{x}_1, \underline{x}_2, ..., \underline{x}_N) = p(\underline{x}_1, ..., \underline{x}_{p-1}, \underline{x}_{p+1}, ..., \underline{x}_N|\underline{x}_p)\,p(\underline{x}_p) \quad [2.110]$$

where $p(\underline{x}_1, ..., \underline{x}_{p-1}, \underline{x}_{p+1}, ..., \underline{x}_N|\underline{x}_p)$ is the conditional probability density. This function represents the probability density for the presence of all scatterers, except the pth scatterer, which is fixed at $\underline{x}_p$. The configurational average with a fixed scatterer is defined by the relation:

$$\langle f(\underline{x}\,|\underline{x}_p)\rangle_{N-1} = \int_V f(\underline{x}\,|\underline{x}_1, ..., \underline{x}_N)p(\underline{x}_1, ..., \underline{x}_{p-1}, \underline{x}_{p+1}, ... \underline{x}_N|\underline{x}_1) \\ \mathrm{d}\underline{x}_1 ... \,\mathrm{d}\underline{x}_{p-1}\,\mathrm{d}\underline{x}_{p+1} ... \,\mathrm{d}\underline{x}_N \quad [2.111]$$

Applying the configurational average [2.109] to relation [2.107] over all possible positions of the scatterers leads to the equation:

$$\begin{aligned}\langle \Phi(\underline{x})\rangle_N &= \Phi_{inc}(\underline{x}) \\ &+ \sum_{m=1}^{N} \int_V T(\underline{x}_n)\Phi_i^{(m)}(\underline{x}, \underline{x}_m)p(\underline{x}_1, ..., \underline{x}_N)\,\mathrm{d}\underline{x}_1 ... \,\mathrm{d}\underline{x}_N\end{aligned} \quad [2.112]$$

Using expressions [2.110] and [2.111], we get:

$$\langle \Phi(\underline{x})\rangle_N = \Phi_{inc}(\underline{x}) + \sum_{m=1}^{N} \int_V T(\underline{x}_m)\left\langle \Phi_i^{(m)}(\underline{x}\,|\underline{x}_m)\right\rangle_{N-1} p(\underline{x}_m)\,\mathrm{d}\underline{x}_m \quad [2.113]$$

with:

$$\left\langle \Phi_i^{(m)}(\underline{x}\,|\underline{x}_m)\right\rangle_{N-1} = \int_V \Phi_i^{(m)}(\underline{x}, \underline{x}_m)p\;\;(\underline{x}_1, ..., \underline{x}_{m-1}, \underline{x}_{m+1}, ..., \underline{x}_N|\underline{x}_m) \\ \mathrm{d}\underline{x}_1\,\mathrm{d}\underline{x}_{m-1}\,\mathrm{d}\underline{x}_{m+1} ... \,\mathrm{d}\underline{x}_N \quad [2.114]$$

The distribution being monodisperse, the probability density $p(\underline{x}_m)$ is related to the number of scatterers per unit volume $n(\underline{x}_m)$ through: $n(\underline{x}_m) = Np(\underline{x}_m)$. Assuming a large number of scatterers, the Foldy–Lax approximation (Foldy 1945):

$$\left\langle \Phi_i^{(m)}(\underline{x}\,|\underline{x}_m)\right\rangle_{N-1} \approx \left\langle \Phi_i^{(m)}(\underline{x}\,,\underline{x}_m)\right\rangle_N \qquad [2.115]$$

translates that the incident field on the fixed scatterer m, averaged over all possible positions of the other scatterers, is approximately equal to the averaged field. According to Twersky, this approximation leads us to neglect all the scattering loops between each pair of scatterers (Ishimaru 1978). Considering the Foldy–Lax approximation [2.115], relation [2.113] takes the form of the Foldy–Twersky equation (Foldy 1945):

$$\langle\Phi(\underline{x}\,)\rangle_N = \Phi_{inc}(\underline{x}\,) + \int_V T(\underline{x}_m)\left\langle \Phi_i^{(m)}(\underline{x}\,,\underline{x}_m)\right\rangle_N n(\underline{x}_m)\,\mathrm{d}\underline{x}_m \qquad [2.116]$$

Assuming a low concentration of localized scatterers, uniformly distributed, the number of scatterers per unit volume corresponds approximately to the average number of scatterers per unit volume n_0 : $n(\underline{x}_m) \approx n_0$ and the transition operator can be expressed as:

$$T(\underline{x}_m)\left\langle \Phi_i^{(m)}(\underline{x}\,,\underline{x}_m)\right\rangle_N = f(\underline{e}_i,\underline{e}_j)\frac{e^{ik|\underline{x}\,-\underline{x}_m|}}{|\underline{x}\,-\underline{x}_m|}\langle\Phi(\underline{x}_m)\rangle_N \qquad [2.117]$$

where $f(\underline{e}_i,\underline{e}_j)$ is the far-field scattering function, calculated for a single scatterer. The vector $\underline{e}_i$ is the unit vector in the propagation direction of $\langle\Phi(\underline{x}_m)\rangle_N$, while vector $\underline{e}_j$ is the unit vector in the direction $\underline{x}\,-\underline{x}_m$.

In the following, the Foldy–Twersky equation is solved in the case of perfectly spherical scatterers of radius R, contained within two parallel planes located at $z = 0$ and $z = d$ (Figure 2.15). The distribution is random and uniform, and the particles do not interpenetrate each other.

The incident wave is a plane wave of unit amplitude, propagating in the z direction:

$$\Phi_{inc}(\underline{x}\,) = \Phi_{inc}(z) = e^{ikz} \qquad [2.118]$$

Considering that the coherent wave also propagates in the z direction, the substitution of expression [2.117] for the transition operator into the Foldy–Twersky equation [2.116] leads to the relation:

$$\langle\Phi(z)\rangle = e^{ikz} + n_0\int_0^h \langle\Phi(z_m)\rangle\left[\iint_{-\infty}^{+\infty} f(\underline{e}_i,\underline{e}_j)\frac{e^{ik|\underline{x}\,-\underline{x}_m|}}{|\underline{x}\,-\underline{x}_m|}\,\mathrm{d}x_m\,\mathrm{d}y_m\right]\mathrm{d}z_m \quad [2.119]$$

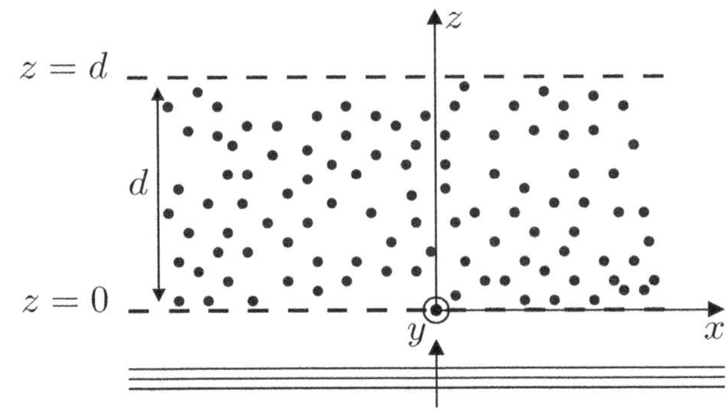

Figure 2.15. *Layer of thickness d containing a random distribution of spherical scatterers*

The double integral between square brackets is calculated by applying the stationary phase method, discussed in section 1.2.2.1, first to the variable x_m and then to the variable y_m :

$$\iint_{-\infty}^{+\infty} f(\underline{e}_i, \underline{e}_j) \frac{e^{ik|\underline{x} - \underline{x}_m|}}{|\underline{x} - \underline{x}_m|} \, \mathrm{d}x_m \, \mathrm{d}y_m \approx i\frac{2\pi}{k} f(\underline{e}_z, \underline{e}_j) e^{ik|z - z_m|} \qquad [2.120]$$

hence:

$$\langle \Phi(z) \rangle = e^{ikz} + i\frac{2n_0\pi}{k} \int_0^h f(\underline{e}_z, \underline{e}_j) \langle \Phi(z_m) \rangle \, e^{ik|z - z_m|} \, \mathrm{d}z_m \qquad [2.121]$$

The vector $\underline{e}_j$ is collinear to e_z and its sign depends on the sign of $z - z_m$. The integral is split into two parts:

$$\begin{aligned} \langle \Phi(z) \rangle = e^{ikz} &+ i\frac{2n_0\pi}{k} e^{ikz} \int_0^z f(\underline{e}_z, \underline{e}_z) \langle \Phi(z_m) \rangle \, e^{-ikz_m} \, \mathrm{d}z_m \\ &+ i\frac{2n_0\pi}{k} e^{-ikz} \int_z^h f(\underline{e}_z, -\underline{e}_z) \langle \Phi(z_m) \rangle \, e^{ikz_m} \, \mathrm{d}z_m \end{aligned} \qquad [2.122]$$

The first two terms of the right-hand side represent the propagation of progressive waves, while the third term represents the propagation of regressive waves. It then follows that the functions $f(\underline{e}_z, \underline{e}_z)$ and $f(\underline{e}_z, -\underline{e}_z)$ correspond to the far-field scattering functions $f_r(0)$ and $f_r(\pi)$, representing, respectively, forward and backward scattering of a longitudinal wave. If the backward scattering is negligible compared to the forward scattering[1], we get:

$$\langle\Phi(z)\rangle = e^{ikz}\left[1 + i\frac{2n_0\pi}{k}f_r(0)\int_0^z \langle\Phi(z_m)\rangle\, e^{-ikz_m}\,\mathrm{d}z_m\right] \qquad [2.123]$$

Upon multiplying this equation by e^{-ikz} and by taking the derivative of the result with respect to z, the integral equation is reduced to a first-order differential equation:

$$\frac{\mathrm{d}\left[\langle\Phi(z)\rangle\, e^{-ikz}\right]}{\mathrm{d}z} = i\frac{2n_0\pi}{k}f_r(0)\,\langle\Phi(z)\rangle\, e^{-ikz} \qquad [2.124]$$

whose solution is a simple plane wave $\langle\Phi(z)\rangle = e^{ik_{eff}z}$, with an effective wave number equal to:

$$k_{eff} = k + \frac{2n_0\pi}{k}f_r(0) \qquad [2.125]$$

In the case of spherical scatterers, and given formula [2.68] with $P_n(1) = 1$ and $k_L = k$, the expression for the effective wave number is:

$$k_{eff} = k - \frac{2in_0\pi}{k^2}\sum_{n=0}^{\infty}(2n+1)T_n \qquad [2.126]$$

where T_n is the scattering coefficient of order n for the longitudinal wave, that is, T_n^{LL}.

In the case of cylindrical scatterers, the same reasoning leads to the relation:

$$k_{eff} = k - \frac{2in_0}{k}\sum_{n=0}^{\infty}\epsilon_n T_n \qquad [2.127]$$

The two previous expressions are approximations of the formulae:

$$\begin{cases} k_{eff}^2 = k^2 - 4in_0\sum_{n=0}^{\infty}\epsilon_n T_n & \text{for cylinders} \\ k_{eff}^2 = k^2 - \frac{4in_0\pi}{k}\sum_{n=0}^{\infty}(2n+1)T_n & \text{for spheres} \end{cases} \qquad [2.128]$$

1 For example, the scattering by a rigid cylinder at high frequencies.

which are often attributed to Foldy, although this author considered point-like scatterers with isotropic radiation. They remain the simplest formulae giving the effective wave number. If backward scattering is not negligible, a reasoning similar to that just described leads to the effective wave number of Waterman and Truell (Waterman and Truell 1961). Moreover, formula [2.128] are still valid for longitudinal waves propagating in solids using the scattering coefficients T_n^{LL} instead of T_n. For shear coherent waves, the formula for cylinders remains valid by using the scattering coefficients T_n^{TT} instead of T_n. However, in the three-dimensional case, due to the T and S polarizations, the effective wave number of coherent shear waves propagating in a dispersion of spherical particles embedded in a solid matrix is given by Tsang and Kong (1982):

$$k_{eff} = k_T \left[1 - \frac{3i\phi}{4(k_T R)^3} \sum_{n=1}^{\infty} (2n+1) \left(T_n^{TT} + T_n^{SS} \right) \right] \qquad [2.129]$$

Equation [2.113] was obtained by fixing a particle and by averaging over all possible positions of the other particles. This mathematical operation can be repeated for several particles, leading to a hierarchy of equations. However, if the number of particles N is large, it is preferable to break up this hierarchy by introducing an approximation. Previously, this hierarchy has been restricted to the first-order by using the *independent scattering approximation* (ISA). Lax (1952) introduced a second-order approximation, by performing the calculation for two fixed particles. This *quasi-crystalline approximation* (QCA) gives rise to several different expressions for the effective wave numbers. In particular, Fikioris and Waterman (1964) take into account the fact that two scatterers cannot interpenetrate by introducing the *hole correction*, which imposes an exclusion distance in the conditional probability density.

Following these works, many formulae, more or less complex, have been proposed to describe the propagation in a medium containing scatterers (Linton and Martin 2005; Conoir and Norris 2010). The advantage of approaches based on the multiple scattering theory lies in the simplicity of the expressions for the effective wave numbers because these only require the determination of scattering coefficients inherent to the problem being studied. However, these formulae are valid as long as the concentration of scatterers and the scattering cross-section remain small.

2.4.2. *Velocity and attenuation of coherent waves*

The properties of the coherent wave that can be measured are its phase velocity and its attenuation, defined by:

$$V_{eff} = \frac{\omega}{\mathrm{Re}\left[k_{eff}\right]} \quad \text{and} \quad \alpha_{eff} = \mathrm{Im}\left[k_{eff}\right] \qquad [2.130]$$

We will study these properties in two cases: a forest of immersed cylindrical rods and an elastic material containing a distribution of resonant spherical particles.

2.4.2.1. *Propagation of coherent acoustic waves in a forest of cylindrical rods*

Coherent waves propagating in a forest of cylindrical steel rods immersed in water were measured using the phase difference method introduced in section 2.4.2.4.2 of Volume 1. The rods, of diameter 1.5 mm, are distributed in a layer of thickness $d = 15$ mm and 20 cm wide. The sample is laterally displaced over thirty positions and transmission measurements are carried out for each position using circular piezoelectric transducers of diameter 13 mm and center frequency 2.25 MHz. The acquired signals are averaged in order to reduce the incoherent field. A reference signal $s_{ref}(t)$, corresponding to the propagation from the emitter to the receiver, without the sample, is also acquired.

In order to observe the influence of averaging on the signal, the average signals s_{avg} and the first recorded signal $s_1(t)$ are shown in Figure 2.16(a). The moduli of the Fourier transforms of the average and reference signals, normalized by the maximum of the modulus of the reference signal spectrum, are shown in Figure 2.16(b). The vertical dashed lines delimit the frequency domain of the study. These limits are imposed to ensure the assumption of plane waves, which are necessary in the phase difference method. The minimum frequency is chosen in order to avoid diffraction by the edges of the sample due to the beam divergence (see section 1.1.4.3), and the maximum frequency is imposed so that the sample is located in the far field.

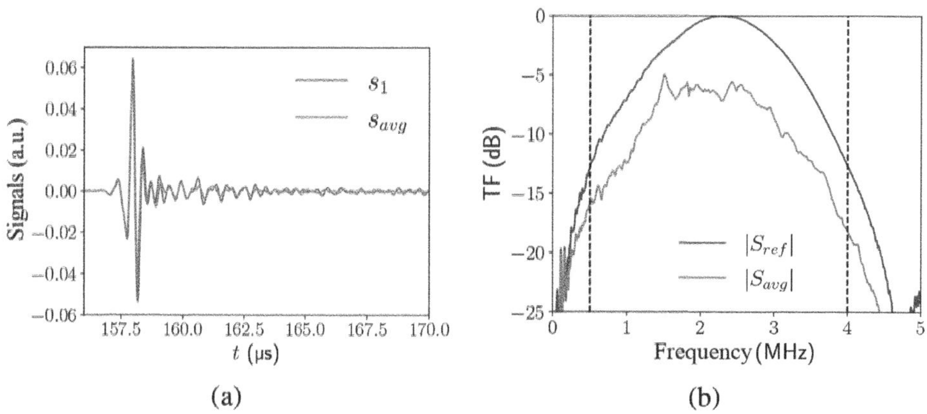

Figure 2.16. *a) Raw signal $s_1(t)$ and averaged signal $s_{avg}(t)$. b) Moduli of the Fourier transforms of the averaged signal ($|S_{avg}|$) and reference signal ($|S_{ref}|$), normalized by the maximum of the modulus of the Fourier transform of the reference signal. For a color version of this figure, see www.iste.co.uk/royer/waves2.zip*

Since there is no physical interface, the phase velocity and attenuation of the coherent wave are given by relations [2.120] in Volume 1, with $h = d$, $\rho = \rho_f$, and $Z_L = Z_f$:

$$V_{eff} = \frac{\omega d}{k_f d - \arg\left[S_{avg}(\omega)/S_{ref}(\omega)\right]} \quad \text{and} \quad \alpha_{eff} = -\frac{1}{d}\ln\left|\frac{S_{avg}(\omega)}{S_{ref}(\omega)}\right| \qquad [2.131]$$

These quantities are plotted in Figure 2.17 as a function of the frequency for a plate containing a random distribution of steel cylinders, of concentration $\phi = 15\%$. The experimental results are compared to expression [2.128], as well as to the Fikioris and Waterman model (1964), which is a second-order approximation with respect to the concentration. Its first-order term is the same as that in relation [2.128]. This comparison shows, for the velocity and attenuation, that the second-order term in concentration must be taken into account in the modeling (Derode *et al.* 2006). These results must be compared with the scattering cross-section of a single steel cylinder immersed in water, as depicted in Figure 2.5. The attenuation and phase velocity are strongly impacted by the propagation of circumferential waves (see section 2.1.3).

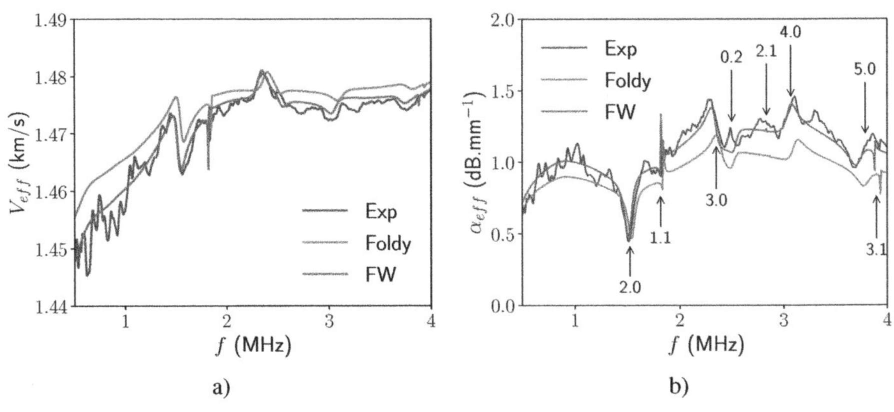

Figure 2.17. *a) Phase velocity and b) attenuation of the coherent wave versus frequency, in a distribution of steel cylinders immersed in water. The arrows indicate the position of the circumferential waves. For a color version of this figure, see www.iste.co.uk/royer/waves2.zip*

2.4.2.2. *Propagation of coherent longitudinal waves in a dispersion of beads*

The velocity and attenuation of the coherent longitudinal wave in a composite material, made up of an epoxy resin containing a distribution of tungsten carbide beads, are plotted in Figure 2.18 *versus* frequency. The distribution is disordered and monodisperse and the concentration of particles is $\phi_0 = 2\%$ (Duranteau *et al.* 2016). The experimental results obtained by a transmission and immersion method (see section 2.4.2.4.2, Volume 1) are compared to the Foldy effective wave number (formula [2.128]), as well as to the LCN multiple scattering model (Luppé *et al.* 2012), which is the generalization of the Fikioris and Waterman model to the elastic case. The translational dipolar resonance of the tungsten carbide beads alters the propagation of the coherent waves: the phase velocity becomes strongly dispersive and an attenuation peak appears in the vicinity of the resonance, whose approximate frequency ($f_1 = 450$ kHz) is given by equation [2.66]. The LCN model is in better agreement with experimental results because it takes into account the conversion into transverse waves, which is particularly large in the vicinity of the resonance, as shown in Figure 2.10(b) for a lead bead.

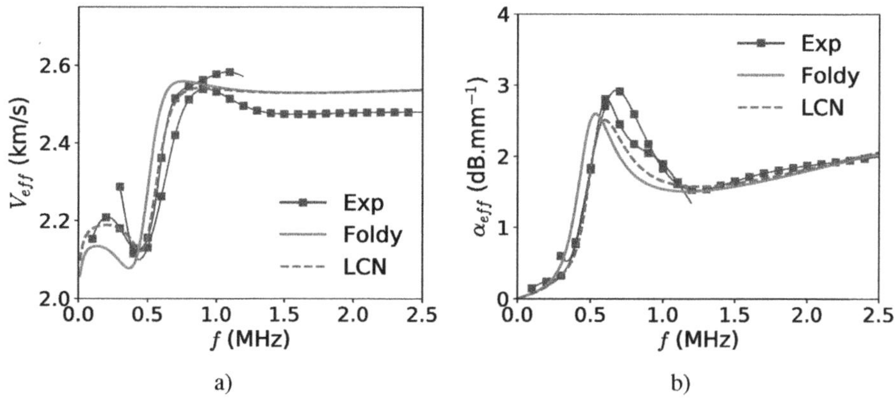

Figure 2.18. *a) Phase velocity and b) attenuation of the coherent longitudinal wave in a composite material, made of an epoxy resin, containing tungsten carbide beads. For a color version of this figure, see www.iste.co.uk/royer/waves2.zip*

3

Generation and Detection

The propagation modes of elastic waves in an unbounded solid, on a free surface or at the interface between two media or in a plate, being known (Volume 1), it is necessary to examine the appropriate means of generating these waves and detecting them efficiently. The techniques implemented are very varied, insofar as they depend on practical constraints and technological possibilities.

If, for example, no mechanical contact with the sample is allowed, one solution could be to generate bulk waves or surface waves using a laser pulse, and to detect them, after propagation, using optical interferometry (Scruby and Drain 1990). At low frequency ($f < 1$ MHz) and when the acoustic impedance Z of the propagation medium is relatively small ($Z < 10$ MRayl), the coupling with the transducer can be achieved through the ambient air. Electromagnetic acoustic transducers (EMAT) are used in the industry to inspect metallic or ferromagnetic materials, by using transverse horizontal or guided waves (e.g. Thompson 1990).

However, the most common and efficient way is to exploit the piezoelectric effect, which couples mechanical and electrical quantities. The mechanism and the conditions of its existence in a crystal (absence of center of symmetry) have been described in section 1.4 of Volume 1. The electromechanical coupling coefficient, which characterizes the efficiency of the conversion, was introduced during the analysis of the propagation of bulk and surface waves. We start by studying the generation and the detection of bulk waves, then of surface waves by the piezoelectric effect, before describing the laser-ultrasonic techniques. Here, we analyze the operation of these devices, knowing that the radiation of sources in a fluid or a solid has been discussed, in the harmonic or impulsive regime, in Chapter 1.

3.1. Piezoelectric transducer for bulk waves

This section deals with the generation (and detection) of longitudinal or transverse bulk waves, using a piezoelectric solid equipped with electrodes, in contact with the propagation medium. The operation of this device, called a transducer, is analyzed from various one-dimensional models, all derived from the same equations but presented in different forms. The piezoelectric solid, which is the active part of the transducer, is first described by an impedance matrix. It corresponds to a hexapole, including an electrical port and two acoustic ports, from which an equivalent electrical circuit is deduced (the classic Mason or Mason–Redwood model). The different elements such as electrodes, impedance-matching layers, rear load (absorbing medium) and front load (propagation medium) are taken into account and represented by quadripoles or simple impedances. Piezoelectric transducers are well described by means of equivalent electrical circuits and the frequency and impulse responses can be predicted in simple cases. We start by specifying, for a single-element transducer and according to the frequency domain, the structures and the main materials used.

3.1.1. *Materials and structures*

The piezoelectric material is the basic element of a transducer since it converts electrical energy coming from an external source into acoustic energy, exploitable in the form of a bulk wave propagating in the medium of study, and vice versa. In its simplest version, the transducer basically comprises a piezoelectric plate, of thickness d and whose lateral dimensions are large with respect to the wavelength λ_P, equipped with electrodes. The electric field, applied across the electrodes, sets the free piezoelectric plate to vibrate at a frequency f_P corresponding to the half-wavelength resonance:

$$d = \frac{\lambda_P}{2} \qquad \longrightarrow \qquad f_P = \frac{V_P}{2d} \qquad [3.1]$$

where V_P is the velocity of elastic waves (longitudinal or transverse) in the piezoelectric material.

The structure of the transducer (Figure 3.1) and the choice of the piezoelectric material depend on the frequency of the waves to be generated and on the propagation medium (solid or liquid). The different media are characterized by their acoustic impedance $Z = \rho V$, which is expressed in MRayl (10^6 Rayl).

3.1.1.1. *Low-frequency domain ($f < 25$ MHz)*

This is, typically, the frequency range for applications in the non-destructive evaluation of materials and in medical ultrasound imaging. The ideal piezoelectric material must have the following properties:

– an acoustic impedance close to that of water or biological tissues: $Z_f \approx$ 1.5 MRayl, with $V_f \approx 1\,500$ m/s and $\rho_f \approx 10^3$ kg/m^3;

– a large electromechanical coupling coefficient K for a better compromise between the sensitivity and the bandwidth of the transducer;

– an electrical impedance close to that of the electronic circuits used in ultrasound imaging;

– a mechanical flexibility that is sufficient to preform the material, for example, giving it a spherical shape, in the case of a focused transducer.

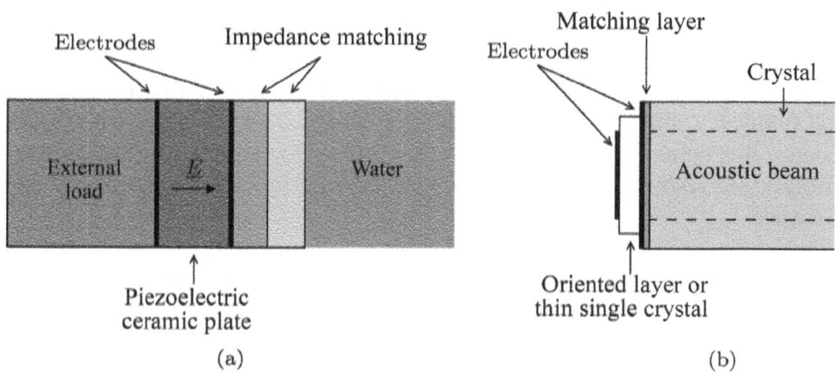

Figure 3.1. *Structure of a single-element plane transducer used to generate bulk waves a) at low frequencies ($f < 25$ MHz) and b) at medium and high frequencies ($f > 25$ MHz)*

Currently, the best solution is provided by a composite material, which consists of a passive polymer phase and many piezoceramic elements forming the active phase. Piezoceramics of the PZT type (lead, zirconium, titanium) and composition $Pb(Zr_xTi_{1-x})O_3$ are synthetic materials made up of ferroelectric grains rigidly bound together by sintering. They become piezoelectric under the action, at high temperature, of an intense electric field, which preferentially aligns the polarization of the grains along its direction. The mechanical, dielectric and piezoelectric properties, very sensitive to the composition of zirconium and titanium, are optimal for $x = 0.52$. These ceramics have both a high electromechanical coupling coefficient ($0.5 \leq K \leq 0.7$) and a large acoustic impedance: $Z_P \geq 30$ MRayl, with $V_P \approx 4\,500$ m/s and $\rho_P \approx 7\,500$ kg/m^3. If the transducer is directly in contact with the propagation medium or through a liquid (water), the elastic wave train generated by the applied electrical pulse will have the shape of a sinusoid, of frequency f_P, with little damping. The axial resolution, equal to several wavelengths, will be poor. It is necessary to absorb the elastic waves emitted backward, by loading the piezoelectric plate with an absorbing and/or scattering resin, often called "backing"

(Figure 3.1(a)). The plate (damped resonator) no longer vibrates, the frequency response is broadened, the impulse response is shorter and the axial resolution is improved. However, the impedance matching between the ceramic and the propagation medium remains a problem. According to the results in section 2.3.1 of Volume 1, the energy transmission coefficient T_{12} between a PZT ceramic and water:

$$T_{12} = 1 - R_{12} = \frac{4Z_P Z_f}{(Z_P + Z_f)^2} \approx 0.17 \qquad [3.2]$$

is only 17%. To increase the efficiency of the electrical energy $\leftrightarrow$ mechanical energy conversion, it is necessary to interpose an intermediary layer, whose thickness is equal to the quarter of the wavelength, and with impedance $Z_I = \sqrt{Z_P Z_f}$ of the order of 7 MRayl (section 2.3.2, Volume 1). With such a large ratio Z_P/Z_f and a single matching layer, the relative bandwidth is limited to 70%. To enlarge this, several frequency shifted layers are required. In practice, these layers, whose acoustic impedance lies between that of a fluid (a few MRayl) and that of a solid (a few tens of MRayl), are resin based. Multiple layers increase the attenuation of the emitted elastic waves.

One way to reduce the number of matching layers is to use a *piezocomposite material*. The transducer consists of ceramic active elements in the form of rods, for example, embedded in a passive resin of much smaller acoustic impedance (3 MRayl). This structure is obtained by cutting a solid ceramic plate in two perpendicular directions and then filling the grooves with the resin. After curing the resin, the base of the plate is mechanically leveled (Figure 3.2).

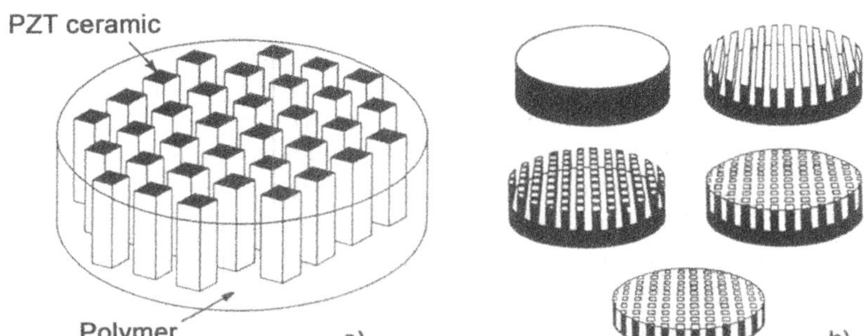

Figure 3.2. *a) Structure of a piezocomposite transducer. b) Manufacturing process steps*

The elastic and piezoelectric constants of the composite depend on those of the ceramic and the polymer and of the volume fraction ϕ of the ceramic (Smith and Auld

1991). The mass density and the permittivity increase linearly with ϕ. It is almost the same for the acoustic impedance, whose value is in between 5 and 15 MRayl for a ceramic volume fraction between 15 and 50%. The coupling coefficient increases very rapidly from $\phi = 0$ (the polymer is passive) to reach a plateau ($20\% < \phi < 80\%$), and then falling toward the value of the piezoceramic plate ($\phi = 1$). Despite appearances, the electromechanical coupling coefficient is larger, over the plateau, than that of a homogeneous plate.

Indeed, for the same applied voltage, a rod surrounded by a relatively soft material (the resin) elongates more than a rod trapped within the bulk of the ceramic. In the multiple-element array, used, for example, in the medical field, the presence of the resin considerably reduces the mechanical coupling between the elements, due to the Lamb waves propagating laterally in the plate (section 4.3, Volume 1). For a typical value $Z_P = 10$ MRayl and a single intermediary layer, the relative bandwidth of the impedance matching to water exceeds 100%.

The piezocomposite transducer, which is widely used, has the disadvantage of an operating range limited to temperatures below 100°C and to frequencies below 20 MHz. Indeed, the period of the rod distribution in the polymer matrix must be small compared to the plate thickness, so that the lateral resonance frequency due to Bragg reflections is higher than the thickness resonance frequency.

3.1.1.2. *Medium frequency domain (25 MHz $< f <$ 500 MHz)*

One technique consists of bonding a thick single crystal ($d > 100$ μm) with a resin or soldering it by metallic diffusion and then thinning it down to the required thickness (between 5 and 100 μm) by polishing. Figure 3.1(b) shows a transducer for acoustic microscope, acousto-optic modulator or high-frequency delay line. Due to its high electromechanical coupling coefficient, lithium niobate is often used for this type of application according to the cuts $Y + 36°$ for longitudinal waves and $Y + 163°$ for transverse waves (section 1.4.3.3, Volume 1). Another type of transducer, based on polymers and copolymers, such as P(VDF-TrFE), composed of vinylidene fluoride (VDF) and trichloroethylene (TrFE), can operate up to frequencies of 100 MHz (Ohigashi *et al.* 1984). These materials have some advantages: their flexibility lends itself to a variety of shapes; their low acoustic impedance avoids the use of any matching layer and their availability in large surfaces leads to applications in medical acoustics and non-destructive testing (acoustic microscopy, film sensors integrated in a composite structure).

3.1.1.3. *High-frequency domain ($f > 1$ GHz)*

The piezoelectric material is deposited in the form of a thin layer over a substrate that is already coated with a metallic film, which constitutes the internal electrode. Most often, this is a layer of zinc oxide (ZnO) or aluminum nitride (AlN), deposited by sputtering on an alumina crystal (sapphire), a propagation medium that has very small losses at high frequencies. As the A_6 axis of ZnO is always perpendicular to the surface, the layer only generates longitudinal waves. AlN is particularly interesting at high frequencies, because of its high velocity; with $V_L = 10\,400$ m/s and $d = 2$ μm: $f_P = 2.6$ GHz. In order to create transverse waves, a plate of lithium niobate with $Y+163°$ cut must be bonded by indium diffusion and brought to the required thickness (a few μm), first by mechanical polishing, then chemical etching or ion bombardment. In this type of transducer, the role of electrodes, at least that of the internal electrode, must be taken into account in the modeling.

The main piezoelectric materials and the useful characteristics for their role as transducer are given in Table 3.1.

Material	Cut	Polarization	V (m/s)	Z (MRayl)	K	$\varepsilon^S/\varepsilon_0$
$LiNbO_3$	$Y+36°$	Quasi-L	7 340	34.5	0.49	38.8
$LiNbO_3$	$Y+163°$	Quasi-T	4 560	21.4	0.62	42.8
PZT4 ceramic	Z	L	4 560	34.2	0.51	633
PZT4 ceramic	Z	T $\perp Z$	2 600	19.5	0.71	735
Piezocomposite	Z	L	3 400	10	0.65	200
P(VDF-TrFE)	Z	L	2 400	4.5	0.30	6.0
ZnO	Z	L	6 400	36.3	0.28	8.8
AlN	Z	L	10 400	34	0.17	8.5

Table 3.1. *Velocity (V), acoustic impedance (Z), electromechanical coupling coefficient (K) and relative permittivity ($\varepsilon^S/\varepsilon_0$) of the main piezoelectric materials used as transducers. The constants of the piezocomposite material are typical values for a volume fraction of ceramic of about 30%*

3.1.2. *One-dimensional model, equivalent circuits*

By hypothesis, the thickness d of the piezoelectric plate, which is large compared to that of the electrodes, is small compared to its lateral dimensions. The electric voltage is applied (or received, if the transducer is a receiver) across these electrodes. In these conditions, a one-dimensional model is sufficient (Figure 3.3). One method of analysis consists of associating to this structure an electromechanical circuit obtained by juxtaposing the equivalent circuits of the different parts, force and particle velocity playing similar roles to electric voltage and intensity.

3.1.2.1. *Impedance matrix*

In the following, the conditions for generating, in the x direction, a single longitudinal or transverse mode, of displacement u, are assumed to be satisfied. For the piezoelectric solid, the constitutive equations [1.229] of Volume 1 can be written without indices as:

$$\sigma = c^E \frac{\partial u}{\partial x} - eE \qquad \text{and} \qquad D = \varepsilon^S E + e\frac{\partial u}{\partial x} \qquad [3.3]$$

where c^E, e and ε^S are the appropriate constants, taking into account the orientation of the crystallographic axes and the type of wave generated. For example, in the case of a piezoelectric ceramic or an aluminum nitride layer of Z-cut and a longitudinal wave: $c^E = C_{33}^E$, $e = e_{33}$ and $\varepsilon^S = \varepsilon_{33}^S$.

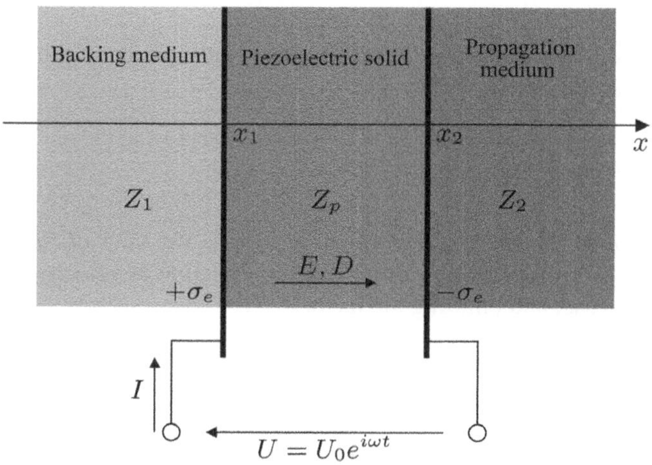

Figure 3.3. *One-dimensional model of a low-frequency transducer, composed of three parts: the propagation medium of impedance Z_2, the piezoelectric material of impedance Z_P and the backing medium of impedance Z_1*

In the piezoelectric solid, assumed to be insulating ($\rho_e = 0$), the Gauss equation ([1.261], Volume 1) imposes that the electric induction D is uniform and, therefore, by continuity (equation [3.108], Volume 1), is equal to the surface density of electric charges σ_e:

$$\frac{\partial D}{\partial x} = \rho_e = 0 \qquad \longrightarrow \qquad D(x,t) = \sigma_e(t) \qquad [3.4]$$

In the harmonic case, the intensity I of the electric current entering through the electrode of abscissa x_1 and with area A is equal to:

$$I(t) = A\frac{\mathrm{d}\sigma_e}{\mathrm{d}t} = i\omega DAe^{i\omega t} \qquad [3.5]$$

The electric field E is eliminated from equations [3.3] in order to express the stress in terms of the electric induction D:

$$\sigma = c^D \frac{\partial u}{\partial x} - hD, \qquad \text{by setting} \qquad h = \frac{e}{\varepsilon^S} \quad \text{and} \quad c^D = c^E + \frac{e^2}{\varepsilon^S} \tag{3.6}$$

where c^D is the stiffness at constant induction. By differentiating with respect to time, it appears the particle velocity $v = \partial u/\partial t$, and current intensity $I = A\partial D/\partial t$:

$$\frac{\partial \sigma}{\partial t} = c^D \frac{\partial v}{\partial x} - \frac{h}{A} I(t) \tag{3.7}$$

Differentiating this equation with respect to x and considering the local equation of dynamics [1.22] of Volume 1, we get:

$$\frac{\partial^2 \sigma}{\partial x \partial t} = c^D \frac{\partial^2 v}{\partial x^2} = \rho \frac{\partial^2 v}{\partial t^2} \tag{3.8}$$

The solution to the propagation equation in v is the sum of the two waves of amplitudes a and b, propagating in the opposite direction at velocity $V = \sqrt{c^D/\rho}$; in the harmonic case, and omitting the factor $e^{i\omega t}$:

$$v(x) = ae^{-ikx} + be^{ikx} \qquad \text{with} \qquad k = \frac{\omega}{V} \tag{3.9}$$

The stress can be deduced from relation [3.7] by introducing this expression for the velocity v:

$$\sigma = -\frac{c^D k}{\omega}\left(ae^{-ikx} - be^{ikx}\right) - \frac{h}{i\omega A} I \tag{3.10}$$

As $c^D = \rho V^2$, the acoustic impedance of the medium appears: $Z = c^D k/\omega = \rho V$.

Let us consider a piezoelectric slab of finite thickness $d = x_2 - x_1$, subject to an electric voltage U applied across the electrodes and to forces F_1 and F_2 exerted by the surrounding material on its faces, of abscissa x_1 and x_2 (Figure 3.4(a)). Since the stress σ corresponds to a traction exerted by the piezoelectric solid, the forces $F = -A\sigma$ exerted on the piezoelectric slab of cross-section A are proportional to the mechanical impedance $\mathcal{Z} = ZA$:

$$F_k = -A\sigma(x_k) = \mathcal{Z}\left(ae^{-ikx_k} - be^{ikx_k}\right) + \frac{h}{i\omega} I, \qquad k = 1, 2 \tag{3.11}$$

In order to express the forces F_1 and F_2 in terms of the velocities on the two faces:

$$v_1 = v(x_1) = ae^{-ikx_1} + be^{ikx_1} \qquad \text{and} \qquad v_2 = -v(x_2) \qquad [3.12]$$

the amplitudes a and b have to be eliminated; after a few lines of calculations, we get:

$$\begin{cases} F_1 = -i\mathcal{Z}\left(\dfrac{v_1}{\tan kd} + \dfrac{v_2}{\sin kd}\right) + \dfrac{h}{i\omega}I \\ F_2 = -i\mathcal{Z}\left(\dfrac{v_1}{\sin kd} + \dfrac{v_2}{\tan kd}\right) + \dfrac{h}{i\omega}I \end{cases} \qquad [3.13]$$

The voltage, U, applied across the electrodes (Figure 3.3) is deduced from the electric field E:

$$U = \int_{x_1}^{x_2} E\,\mathrm{d}x \qquad \text{with} \qquad E = -h\frac{\partial u}{\partial x} + \frac{D}{\varepsilon^S} \qquad [3.14]$$

that is, in terms of the current intensity $I = i\omega AD$ and particle velocities $v_1 = i\omega u(x_1)$ and $v_2 = -i\omega u(x_2)$:

$$U = \frac{h}{i\omega}(v_1 + v_2) + \frac{I}{i\omega C_0} \qquad \text{where} \qquad C_0 = \frac{\varepsilon^S A}{d} \qquad [3.15]$$

is the static capacitance of the rigidly bonded transducer. Equations [3.13] and [3.15] take the form of the following matrix:

$$\begin{pmatrix} F_1 \\ F_2 \\ U \end{pmatrix} = -i \begin{pmatrix} \mathcal{Z}/\tan kd & \mathcal{Z}/\sin kd & h/\omega \\ \mathcal{Z}/\sin kd & \mathcal{Z}/\tan kd & h/\omega \\ h/\omega & h/\omega & 1/C_0\omega \end{pmatrix} \begin{pmatrix} v_1 \\ v_2 \\ I \end{pmatrix} \qquad [3.16]$$

which reveals the *electromechanical impedance matrix* $[\mathcal{Z}]$ of the three-port circuit, shown in Figure 3.4(b). This matrix is symmetric because acoustic ports 1 and 2 are equivalent.

3.1.2.2. *Equivalent circuits*

An *infinite* non-piezoelectric medium, with cross-section A, in which a progressive wave propagates (in the direction of increasing x), is represented by its mechanical impedance $\mathcal{Z} = \rho VA$, that is, the ratio of the force $F = -A\sigma$ to the particle velocity v.

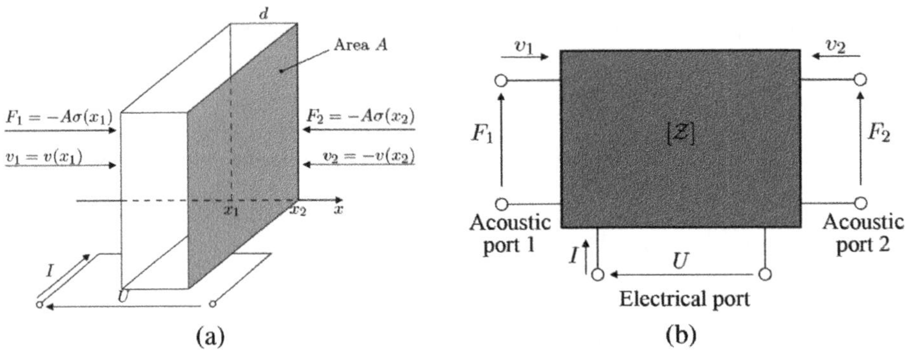

Figure 3.4. *Piezoelectric plate of cross-section A. a) Mechanical quantities: forces and velocities, and electrical quantities: voltage and current intensity. b) Hexapole with two acoustic ports and one electrical port representing a piezoelectric slab*

The equivalent circuit of a (non-piezoelectric) slab of *finite thickness* d and cross-section A can be deduced from relations [3.13] with $h = 0$. Taking into account the identity $1/\tan kd = (1/\sin kd) - \tan(kd/2)$, we get:

$$F_k = -i\frac{\mathcal{Z}}{\sin kd}(v_1 + v_2) + i\mathcal{Z}\tan\left(\frac{kd}{2}\right)v_k, \qquad k = 1, 2 \qquad [3.17]$$

These relations can also be obtained by applying the Kirchhoff laws to the circuit in Figure 3.5(a). The coaxial line element of length d and characteristic impedance $\mathcal{Z}$ in Figure 3.5(b), which obeys the same laws, is also an equivalent circuit for the solid slab.

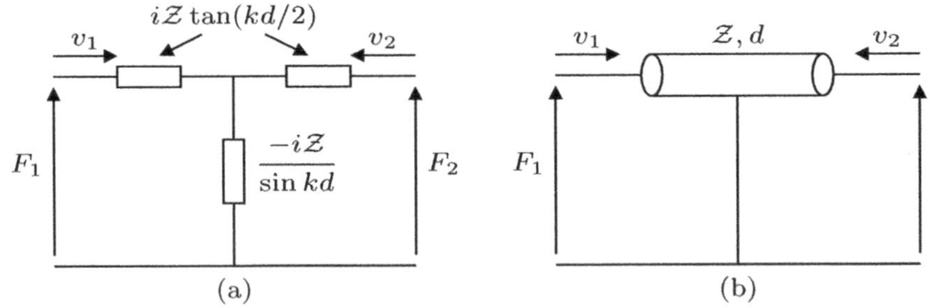

Figure 3.5. *Equivalent circuit to a non-piezoelectric solid slab of thickness d and mechanical impedance $\mathcal{Z}$: a) equivalent T element; b) transmission line element*

When the *material is piezoelectric*, there exists an additional force f in the right-hand sides of equations [3.13]:

$$f = \frac{h}{i\omega} I = hC_0 \left[U - \frac{h}{i\omega}(v_1 + v_2) \right] \qquad [3.18]$$

This force must be introduced into the central branch of the circuits in Figure 3.5, through which passes the velocity $v_1 + v_2$. This is obtained by means of an electromechnical transformer of ratio hC_0 and with a negative capacity $-C_0$ arranged in series in the primary circuit (Figure 3.6), such that the potential difference at the terminals of the primary winding is equal to the term between square brackets in expression [3.18], that is, to f/hC_0.

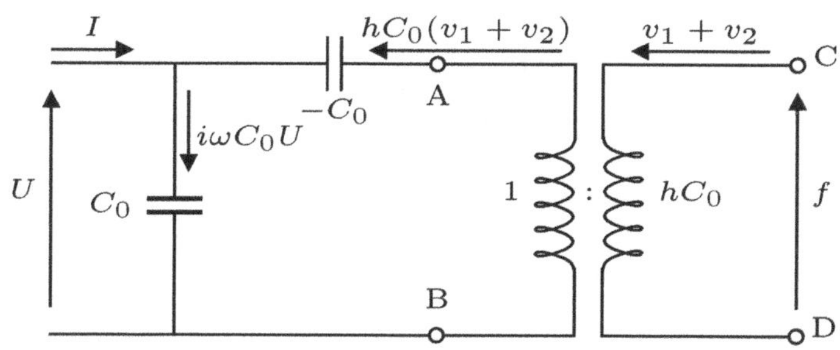

Figure 3.6. *Electromechanical transformer representing the piezoelectricity*

The Mason circuit (Mason 1948), equivalent to a slab of piezoelectric material, is represented in Figure 3.7. It includes an electrical port and two acoustic ports.

The equivalent circuit for the transducer is obtained by associating in cascade the circuits relative to each part. Figure 3.8 represents the electromechanical circuit of a transducer composed of a piezoelectric plate (P), a backing medium (1) and a matching layer (2). The backing being free, the corresponding terminals are short-circuited because the acoustic impedance of the air is very small. The propagation medium, assumed to be infinite, is represented by its mechanical impedance.

In Redwood circuit (Redwood 1961), the T element is replaced by the portion of transmission line in Figure 3.5(b). This section can be considered as a coaxial cable whose external conductor is connected to the secondary of the transformer.

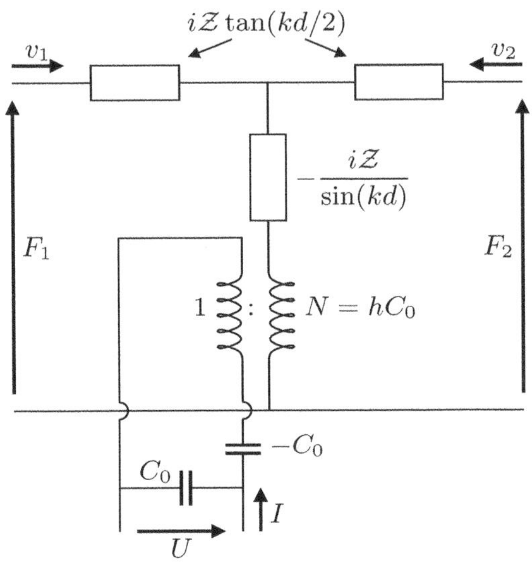

Figure 3.7. *Electromechanical circuit (Mason circuit) equivalent to a slab of piezoelectric material of thickness d, cross-section A and mechanical impedance $\mathcal{Z} = ZA$*

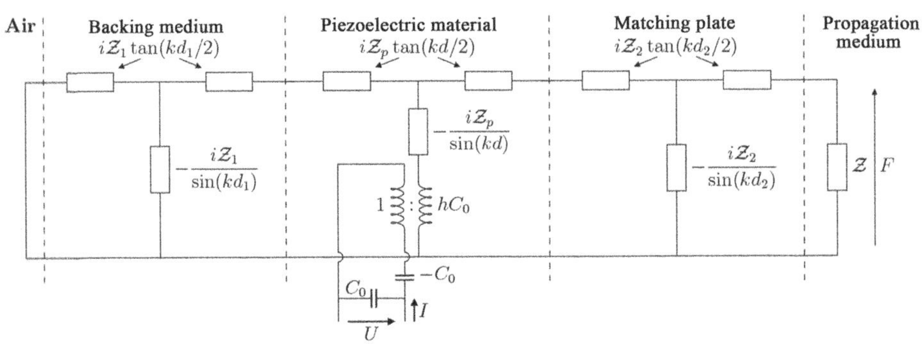

Figure 3.8. *Complete equivalent circuit of a transducer. The left end, in contact with the air, is short-circuited, and the right end is loaded by the mechanical impedance $\mathcal{Z}$ of the propagation medium*

Another equivalent circuit is composed of a transmission line expressing the propagation and an electromechanical transformer modeling the energy conversion. The particle velocity, this time, is injected in the middle of the line. This is the KLM model, proposed by Krimholtz *et al.* (1970).

3.1.3. *Frequency and impulse responses – emitted power*

In its simplest version (Figure 3.3), a single-element bulk wave transducer comprises a piezoelectric plate of thickness d, cross-section A, and mechanical impedance $\mathcal{Z}_P = AZ_P$. The rear face of the plate is bonded to an absorbent backing medium, equivalent to its acoustic impedance Z_1, since no wave is reflected. The other face of the plate is rigidly bonded to the propagation medium, assumed to be infinite and of impedance Z_2. By substituting the forces on both faces, $F_1 = -\mathcal{Z}_1 v_1$ and $F_2 = -\mathcal{Z}_2 v_2$, into the system [3.16], we get:

$$\begin{pmatrix} i\mathcal{Z}_1 + \mathcal{Z}_P/\tan kd & \mathcal{Z}_P/\sin kd & h/\omega \\ \mathcal{Z}_P/\sin kd & i\mathcal{Z}_2 + \mathcal{Z}_P/\tan kd & h/\omega \\ h/\omega & h/\omega & 1/C_0\omega \end{pmatrix} \begin{pmatrix} v_1 \\ v_2 \\ I \end{pmatrix} = \begin{pmatrix} 0 \\ 0 \\ iU \end{pmatrix} \qquad [3.19]$$

The inversion of this system leads to the equation:

$$\begin{pmatrix} v_1 \\ v_2 \\ I \end{pmatrix} = \begin{pmatrix} y_{11} & y_{12} & y_{13} \\ y_{12} & y_{22} & y_{23} \\ y_{13} & y_{23} & y_{33} \end{pmatrix} \begin{pmatrix} 0 \\ 0 \\ iU \end{pmatrix} \qquad [3.20]$$

where y_{ij} are the components of the (symmetric) admittance matrix. The particle velocity $v = -v_2$ in the propagation medium is then expressed by:

$$v = -v_2 = -iy_{23}U \quad \text{with} \quad y_{23} = -\frac{\Delta_{23}}{\Delta} \quad \longrightarrow \quad v = i\frac{\Delta_{23}}{\Delta}U \qquad [3.21]$$

as well as the electrical impedance $Z_e = U/I$:

$$I = iy_{33}U \qquad \text{with} \qquad y_{33} = \frac{\Delta_{33}}{\Delta} \qquad \longrightarrow \qquad Z_e = \frac{\Delta}{i\Delta_{33}} \qquad [3.22]$$

where Δ is the determinant of the impedance matrix [3.20], and Δ_{23} and Δ_{33} are the minors of the components z_{23} and z_{33} of this matrix. By developing with respect to the last line of the impedance matrix, we get:

$$\Delta = \frac{1}{C_0\omega}\left[\Delta_{33} + hC_0(\Delta_{13} - \Delta_{23})\right] \qquad [3.23]$$

with:

$$\begin{cases} \Delta_{13} = \dfrac{h}{\omega}\left[\mathcal{Z}_P \tan\theta - i\mathcal{Z}_2\right] \\ \Delta_{23} = \dfrac{h}{\omega}\left[i\mathcal{Z}_1 - \mathcal{Z}_P \tan\theta\right] \\ \Delta_{33} = i(\mathcal{Z}_1 + \mathcal{Z}_2)\dfrac{\mathcal{Z}_P}{\tan 2\theta} - \mathcal{Z}_1\mathcal{Z}_2 - \mathcal{Z}_P^2 \end{cases} \qquad [3.24]$$

by setting:

$$\theta = \frac{kd}{2} = \frac{\pi f}{2f_P} \quad \text{with} \quad f_P = \frac{V_P}{2d} \qquad [3.25]$$

By introducing the square of the electromechanical coupling coefficient:

$$K^2 = \frac{e^2}{\varepsilon^S c^D} \quad \text{in the term} \quad \frac{h^2 C_0}{\omega} = \frac{e^2}{\varepsilon^S}\frac{A}{kdV_P} = \frac{K^2}{2\theta}\mathcal{Z}_P \qquad [3.26]$$

the particle velocity v is given by:

$$v = \frac{hC_0\left[i\mathcal{Z}_1 - \mathcal{Z}_P \tan\theta\right]}{(\mathcal{Z}_1 + \mathcal{Z}_2)\mathcal{Z}_P\left[\cot(2\theta) - K^2/2\theta\right] + i\left[\mathcal{Z}_1\mathcal{Z}_2 + \mathcal{Z}_P^2 - \mathcal{Z}_P^2 K^2(\tan\theta)/\theta\right]} \qquad [3.27]$$

Since $h = e/\varepsilon^S$ and $C_0 = \varepsilon^S A/d$, the particle velocity can be written in terms of the ratios $r_1 = \mathcal{Z}_1/\mathcal{Z}_P = Z_1/Z_P$ and $r_2 = Z_2/Z_P$ of the acoustic impedances:

$$v = v_0 H(r_1, r_2, \theta)U \quad \text{with} \quad v_0 = \frac{e}{Z_P d} \qquad [3.28]$$

It depends on the relative frequency f/f_P through the intermediary of the variable θ and the dimensionless function:

$$H(r_1, r_2, \theta) = \frac{r_1 + i\tan\theta}{1 + r_1 r_2 - K^2\tan\theta/\theta + i(r_1 + r_2)\left(K^2/2\theta - \cot 2\theta\right)} \qquad [3.29]$$

3.1.3.1. *Frequency response*

The values of the frequency response $H(f)$ of the transducer are close to unity at the center frequency f_P ($\theta = \pi/2$) and when the three impedances are equal ($r_1 = r_2 = 1$): $H(1, 1, \pi/2) \approx 1$. The particle velocity [3.28] is therefore of the order of the quantity $v_0 = e/Z_P d$, which depends only on the piezoelectric material. For a PZT-type ceramic ($e_{33} \approx 20$ C/m^2), whose characteristics are given in Table 3.1, and at a frequency $f_P = 3$ MHz ($d = 0.76$ mm), v_0 is of the order of 10^{-3} m/s per volt. The particle velocity is thus equal to 10 cm/s for a sinusoidal voltage of amplitude $U = 100$ V. This would be larger in the case of a piezocomposite material with a smaller impedance Z_P. Under the same conditions, the amplitude of the mechanical displacement $u_0 = v_0/\omega_P \approx e/\pi c^D$ is equal to 0.05 nm/V, that is, 5 nm for a harmonic voltage of amplitude 100 V.

For example, the transducer is made up of a piezoelectric plate ($K = 0.5$) bonded to a perfectly absorbent backing, with an impedance $Z_P/4$ ($r_1 = 0.25$). The curves in Figure 3.9 are plotted, with respect to the relative frequency f/f_P, for several values of the impedance ratio $r_2 = Z_2/Z_P$. When the impedance Z_2 of the propagation medium is smaller than that of the piezoelectric material ($r_2 = 0.25$ and 0.5), the plate vibrates preferentially at a half-wavelength ($d = \lambda_P/2$). The acoustic power emitted is maximum for $f = f_P$. When Z_2 is larger than Z_P ($r_2 = 2$), the vibrations of the plate, whose two faces are loaded, are highly damped. The frequency response is nearly flat, from $f_P/4$ to $3f_P/4$, and the relative bandwidth exceeds 100%.

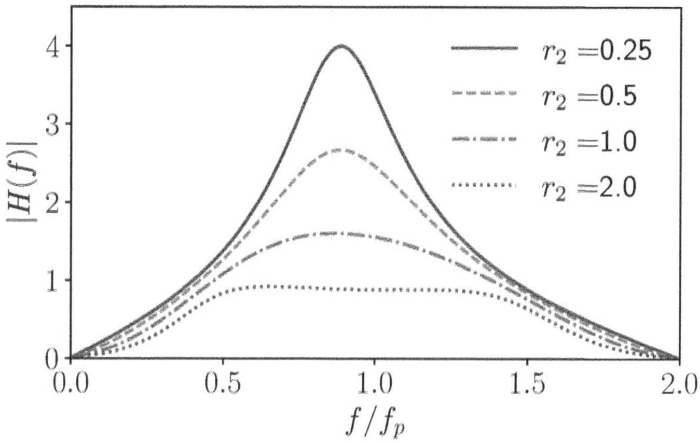

Figure 3.9. *Amplitude of the frequency response $H(f)$ of a transducer ($K = 0.5$), of impedance Z_P, for an absorbent backing medium, of impedance $Z_1 = Z_P/4$, and different impedances $Z_2 = r_2 Z_P$ of the propagation medium. For a color version of this figure, see www.iste.co.uk/royer/waves2.zip*

3.1.3.2. *Time responses*

In the field of non-destructive testing of materials, and in medical ultrasound imaging, transducers operate in impulse regime. The axial resolution is related to the duration of the emitted acoustic wave train, which depends on the transducer bandwidth. The impulse response $h(t)$ is the Fourier transform of the frequency response $H(f)$.

The change in variable $\zeta = f/f_P$ leads to:

$$h(t) = f_P \int_{-\infty}^{+\infty} H(\zeta) e^{-i2\pi\zeta f_P t} \, d\zeta = f_P h_N(f_P t) \qquad [3.30]$$

where $h_N(f_P t)$ is the normalized (dimensionless) impulse response. This one is plotted in Figure 3.10 for the four previous values of the acoustic impedance ratio r_2.

When $Z_2 = Z_P$, the acoustic pulse is very short and almost bipolar (curve (c)). When the transducer resonates in the vicinity of the central frequency f_P (curves (a) and (b)), the particle velocity is larger and the oscillations are less damped. The response (d) ($r_2 = 2$) shows small oscillations at the frequency $3f_P/2$, arising from a slight resonance at $3\lambda_P/4$.

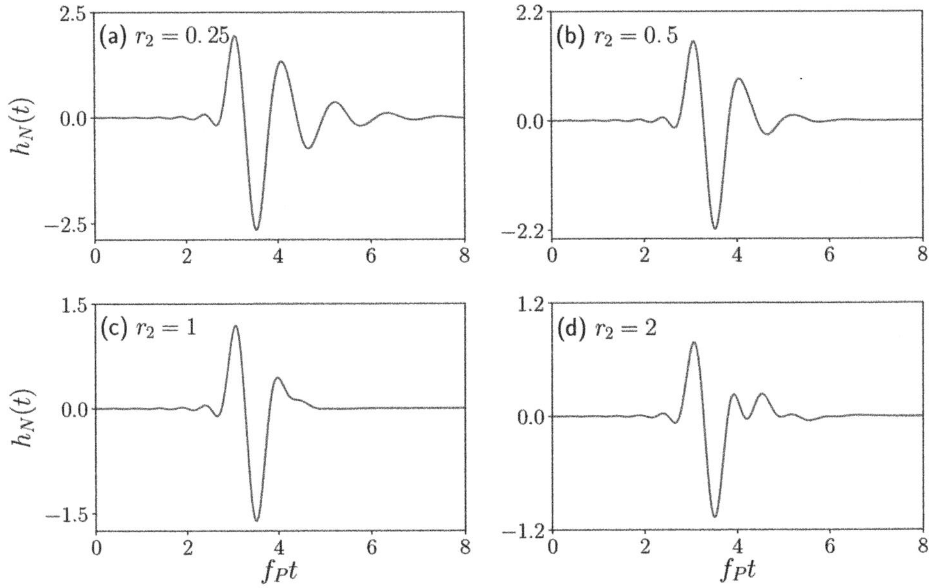

Figure 3.10. *Normalized impulse responses $h_N(f_P t)$ of a PZT ceramic transducer, of impedance Z_P, for a backing medium of impedance $Z_P/4$ and a propagation medium of impedance $r_2 Z_P$ with $r_2 = 0.25; 0.5; 1; 2$*

The particle velocity $v(t)$ is equal to the convolution product between the impulse response $h(t)$ and the electric voltage $U(t)$ applied to the transducer. Using the relation [3.28], established in the harmonic case, $v(t)$ is given by:

$$v(t) = v_0 h(t) \otimes U(t) \qquad [3.31]$$

The electric impulse is, generally, a rectangle of height U_m and duration Θ, short compared to $1/f_P$. The transducer's response is therefore similar to the impulse response, with a multiplicative coefficient $U_m\Theta$ equal to the area of the pulse. By introducing the normalized impulse response, we get:

$$v(t) = v_0 U_m \Theta f_P h_N(f_P t) \qquad \text{if} \qquad \Theta f_P \leq 0,3 \qquad [3.32]$$

ORDER OF MAGNITUDE.– The amplitude of the particle velocity is proportional to the pulse width Θ. However, due to the convolution, the duration of the acoustic signal increases if the condition $\Theta f_P << 1$ is no longer satisfied; a good compromise is $\Theta f_P \approx 0.3$. With $v_0 = 10^{-3}$ ms^{-1}V^{-1} and, since the maximum value of h_N is close to 1.5 (Figure 3.10 for $r_2 = 1$), the peak velocity v_m reaches 0.14 m/s for a voltage $U_m = 300$ V.

REMARK.–

– In an impulse regime, it is possible to apply much higher voltages than in the sinusoidal case because, if the cycle rate is reduced, the heating of the transducer is limited. Moreover, the electric pulse can be produced by the simple discharge of a capacitor.

– The particle velocity $v(t)$, expressed in terms of the voltage $U(t)$ applied to the transducer, may serve as the input quantity for the models discussed in section 1.1.4.1 to calculate the acoustic field radiated in a fluid by a vibrating transducer, in piston mode, for example. It was seen, in this section, that the velocity potential at point M is given by the convolution of the particle velocity $v(t)$ at the front face of the transducer and the diffraction impulse response $h(M, t)$.

3.1.3.3. *Emitted acoustic power*

In the sinusoidal case, the instantaneous power $P(t) = F_2 v$ with $F_2 = -AZ_2 v_2 = AZ_2 v$ has the average value:

$$\langle P \rangle = \frac{1}{2} AZ_2 |v|^2 = \frac{1}{2} Ar_2 \frac{e^2}{Z_P d^2} |H(f)|^2 |U|^2 \qquad [3.33]$$

Given the relations $C_0 = \varepsilon^S A/d$ and $Z_P = c^D/V_P$, it still appears the square of the electromechanical coupling coefficient [3.26] and the resonance frequency of the free plate, $f_P = V_P/2d$. The emitted power, proportional to K^2:

$$\langle P \rangle = K^2 C_0 f_P r_2 |H(f)|^2 |U|^2 = \frac{1}{2} K^2 \varepsilon^S V_P \frac{A}{d^2} r_2 |H(f)|^2 |U|^2 \qquad [3.34]$$

depends on the ratio of the acoustic impedances r_1 and r_2 through the intermediary of the frequency response $H(f)$. The material is involved through the figure of merit $M = K^2 \varepsilon^S V_P$. The dimensionless ratio A/d^2, which characterizes the transducer geometry, is always very large because the lateral dimensions of the piezoelectric plate are greater than 10 times its thickness, in order to preferentially excite the thickness mode of vibration.

ORDER OF MAGNITUDE.– For a PZT ceramic disk of thickness $d = 0.76$ mm ($f_P = 3$ MHz), radius $a = 10$ mm: $A/d^2 = 544$ and as $|H(f_P)| \approx \sqrt{2}$ when $r_2 = 1$ (Figure 3.9), the acoustic power emitted is 3.5 mW/V^2, that is, 9 W for an applied sinusoidal voltage of amplitude 50 V.

3.1.4. *Electrical impedance, efficiency*

Let us calculate the electrical impedance of the transducer using the formula [3.22]:

$$Z_e = \frac{\Delta}{i\Delta_{33}} = \frac{1}{iC_0\omega}\left(1 + hC_0\frac{\Delta_{13} - \Delta_{23}}{\Delta_{33}}\right) \quad [3.35]$$

Given the expressions [3.23] and [3.24] of the determinant Δ and the minors Δ_{ij} of the impedance matrix, and considering relation [3.26], we get:

$$Z_e = \frac{1}{iC_0\omega}\left(1 + \frac{K^2}{2\theta}\frac{2(1 - \cos 2\theta) - i(r_1 + r_2)\sin 2\theta}{i(r_1 + r_2)\cos 2\theta - (1 + r_1 r_2)\sin 2\theta}\right) \quad [3.36]$$

The electrical impedance is complex:

$$Z_e = \frac{1}{iC_0\omega} + iX_a(\omega) + R_a(\omega) \quad \text{or} \quad Y = iC_P\omega + iB_a(\omega) + G_a(\omega) \quad [3.37]$$

In addition to the reactance due to the static capacitance, it includes a real part $R_a(\omega)$ and an imaginary part $X_a(\omega)$ (the motional impedance), which are both frequency dependent (Figure 3.11(a)). $B_a(\omega)$ and $G_a(\omega)$ are the corresponding elements of the parallel equivalent electric circuit (Figure 3.11(b)). The radiation resistance $R_a(\omega)$ and conductance $G_a(\omega)$ reflect the conversion of a part of the electric power supplied by the source into mechanical power, transported by the emitted elastic waves, whose average value is:

$$\langle P\rangle = \frac{1}{2}R_a(\omega)|I|^2 \quad \text{or} \quad \langle P\rangle = \frac{1}{2}G_a(\omega)|U|^2 \quad [3.38]$$

Since the motional reactance $X_a(\omega_P)$ is zero at the center frequency f_P, for which $\theta = \pi/2$, the electrical impedance (equation [3.36]) is composed of a resistance in series with the static capacitance C_0:

$$Z_e(\omega_P) = \frac{1}{iC_0\omega_P} + R_a(\omega_P) \quad \text{with} \quad R_a(\omega_P) = \frac{4K^2}{\pi C_0\omega_P}\frac{Z_P}{Z_1 + Z_2} \quad [3.39]$$

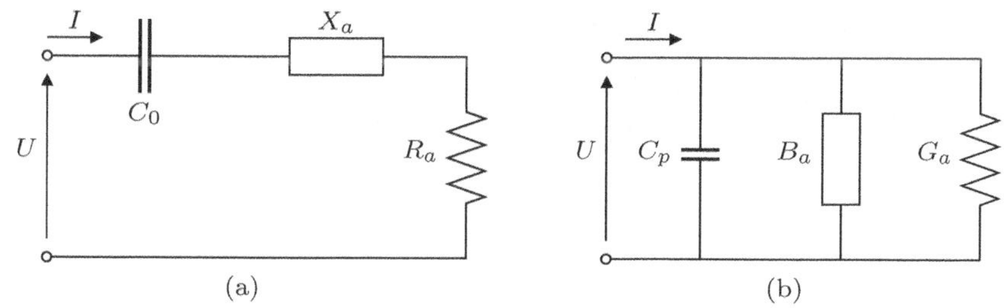

Figure 3.11. *Electrical impedance of a bulk wave transducer. Equivalent circuit: a) series and b) parallel*

In the sinusoidal case, if the transducer is supplied by a generator of electromotive force U_0 and internal resistance R_0 (Figure 3.12(a)), the average power emitted $\langle P \rangle$ is a fraction of the maximum available electrical power P_d at the generator output:

$$\langle P \rangle = \frac{1}{2} \frac{R_a U_0^2}{(R_0 + R_a)^2 + (1/C_0 \omega_P)^2} \quad \text{and} \quad P_d = \frac{U_0^2}{8R_0} \qquad [3.40]$$

The *transducer efficiency* η, defined as the ratio of these two powers:

$$\eta = \frac{\langle P \rangle}{P_d} = \frac{4R_0 R_a}{(R_0 + R_a)^2 + (1/C_0 \omega_P)^2} = \eta(R_0) \qquad [3.41]$$

is maximum ($\eta = \eta_{max}$) for a value R_0, such that $\partial \eta / \partial R_0 = 0$:

$$R_0^2 = R_a^2 + \frac{1}{(C_0 \omega_P)^2} \quad \text{and} \quad \eta_{max} = \frac{2R_a}{R_0 + R_a} \qquad [3.42]$$

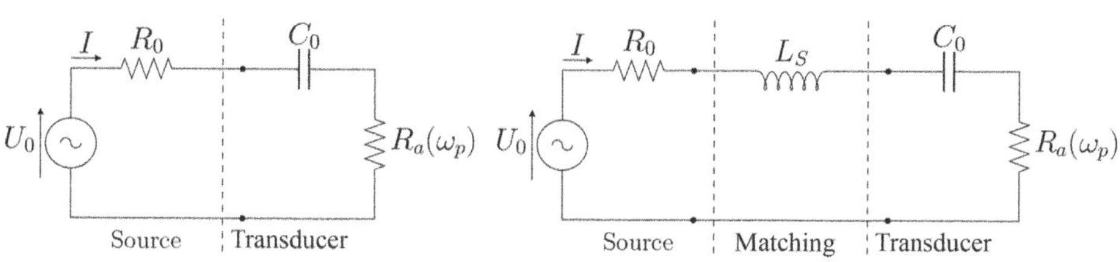

Figure 3.12. *a) Equivalent electrical circuit of a transducer at its center frequency; b) impedance matching through a series inductance L_S*

Let us examine different possibilities:

a) transducer with a backing medium ($Z_1 = Z_P$) and a perfectly matching layer ($Z_2 = Z_P$). If $K \leq 0.5$, the radiation resistance [3.39]:

$$R_a(\omega_P) = \frac{2K^2}{\pi C_0 \omega_P} \leq \frac{1}{2\pi C_0 \omega_P} \qquad [3.43]$$

is much smaller than the reactance $1/C_0\omega_P$ due to the static capacitance and the optimal value of the source resistance R_0 is close to $1/C_0\omega_P$. In practice, for a given material and frequency, this condition determines the static capacitance and, therefore, the area A of the electrodes. The maximum value of the transducer efficiency:

$$\eta_{max} = 2R_0R_a(C_0\omega_P)^2 \approx 2R_aC_0\omega_P \approx \frac{4K^2}{\pi} \qquad [3.44]$$

is only 32% for $K = 0.5$ (PZT ceramic), but the bandwidth is wide;

b) transducer without "backing": $Z_1 = 0$ but with a matching plate ($Z_2 = Z_P$). The radiation resistance and the efficiency are doubled: all the power is emitted into the propagation medium, while in the previous case, half the energy was dissipated into the backing medium. However, the transducer is less well damped: the efficiency increases at the expense of the bandwidth;

c) if the electromechanical coupling coefficient is small, for example, $K \leq 0.2$ (piezoelectric polymer), the maximum efficiency $\eta_{max} = 8K^2/\pi$ is less than 10%. It may be increased by compensating the static capacitance C_0 with an inductance L_S arranged in series (Figure 3.12(b)) such that $L_SC_0\omega_P^2 = 1$. The current intensity, the average power emitted, and the efficiency are equal to:

$$I(\omega_P) = \frac{U_0}{R_0 + R_a}, \quad \langle P \rangle = \frac{1}{2}\frac{R_aU_0^2}{(R_0 + R_a)^2} \quad \text{and} \quad \eta = \frac{4R_0R_a}{(R_0 + R_a)^2} \qquad [3.45]$$

Since $K^2 \leq 1$, R_a is very small compared to R_0:

$$\eta \approx \frac{4R_a}{R_0} = \frac{16K^2}{\pi R_0 C_0 \omega_P} = \frac{16K^2}{\pi}Q, \qquad \text{where} \qquad Q = \frac{1}{R_0C_0\omega_P} \qquad [3.46]$$

is the quality factor of the resonant circuit composed of R_0, L_S, C_0. The efficiency is multiplied by $2Q$, but the relative bandwidth is then less than $1/2Q$. When R_0 and R_a are very different, the electrical matching requires a transformer.

3.2. Piezoelectric transducer for surface waves

This section deals with the generation and detection of surface acoustic waves (SAW) by interdigital electrode transducers, placed on a piezoelectric substrate, and achieved through a technology compatible with that of microelectronics. The piezoelectricity ensures the direct and inverse electromechanical conversion. The transducer structure determines the distribution of the sources and thus the desired function. The accessibility at the free surface and the high concentration of acoustic energy in a thin subsurfacic layer favor the interaction of the SAW with the external environment and are at the origin of the development, these last decades, of numerous applications in various fields. These surface wave components are used as radio-frequency filters for telecommunications, sensors to measure physical or biochemical quantities, acousto-optic modulators and actuators for the microfluidic. The substrate, most often a single crystal, is of prime importance, since it simultaneously plays the role of generation medium and propagation medium.

The piezoelectric materials used for Rayleigh waves were described in section 3.1.3.5 of Volume 1 and their mechanical, piezoelectric and electrical characteristics are presented in Table 3.1. In order to satisfy the demand (arising mainly from the cell phone industry) for components operating at a center frequency higher than 3 GHz, without increasing technological difficulties ($f = 3$ GHz and $V_R = 3\,000$ m/s $\rightarrow \lambda/4 = 0.25\ \mu$m), research was carried out into surface waves with higher velocities. Given the limited number of piezoelectric solids, a first solution is to use a heterogeneous structure composed of a non-piezoelectric substrate, characterized by a high velocity (alumina, diamond in the form of a membrane), carrying a piezoelectric layer (ZnO, AlN). Another solution is to find, in a piezoelectric crystal, cuts able to generate and support pseudo-surface acoustic waves (PSAW) propagating at a velocity greater than that of Rayleigh waves. These PSAW, with a dominant transverse horizontal or even longitudinal component (section 3.1.2.4, Volume 1), are strongly coupled to the interdigital transducer ($K^2 = 20\%$). They are sensitive to the thickness of the metal, which retains them more or less close to the surface, and propagate with very small losses. Thus, it is necessary to add Table 3.2 to Table 3.1 of Volume 1.

Crystal	Cut	Propagation	Velocity (m/s)	K^2 (%)	$\varepsilon_p/\varepsilon_0$	$\frac{1}{\tau}\frac{d\tau}{d\theta}$. ($10^{-6}$ K^{-1})
$LiNbO_3$	$Y + 41°$	X	4 840	21	63	-90
$LiNbO_3$	$Y + 64°$	X	4 695	11.6	52	-90
$LiTaO_3$	$Y + 36°$	X	4 240	4.7	50	-35

Table 3.2. *Characteristics of piezoelectric pseudo-surface acoustic waves*

We begin by examining the operating principle of the interdigital electrode transducer, and then we analyze it according to different models. The first SAW devices used for signal processing were designed by representing the transducer as an ideal transversal filter, whose impulse response was the replica of the comb shape. The advantage of this model is that it can be easily inverted to determine the weighting function leading to the required frequency response. Its main weakness is that it does not take into account internal reflections on the electrodes. Solutions have therefore been proposed to reduce these effects, considered as secondary. The discrete source model (section 3.2.1.2) and the more advanced impulse response model (section 3.2.2) are applied to transducers equipped with electrodes having an appropriate geometry to reduce internal reflections. Both models lead to a finite impulse response.

Piezoelectric regeneration reveals that a receiver transducer, electrically loaded, behaves partially as a transmitter. Any incident wave induces an electric voltage across the load and this voltage in turn generates two waves, which leave the transducer on either side. One of them goes toward the emitter; in turn, this transducer emits a new wave, and so on. The result is a series of spurious echoes, the most important of which is the triple-transit signal (section 3.2.3). The other reason why a wave traveling under an interdigital transducer – loaded or not – generates a wave in the opposite direction is the reflection on the sequence of fingers. This observation is valid for any wave that originates under the active fingers of the transducer. The coupled mode method lends itself to the analysis of a passive array, that is, a reflector, as well as an active transducer (section 3.2.4).

3.2.1. *Operating principle*

The transducer consists of two comb-shaped electrodes, placed on a piezoelectric substrate (for example, by photo-etching a vacuum-deposited metallic film). In its simplest version (Figure 3.13(a)), the space d between the fingers of the comb and the overlap width w of two adjacent fingers are constant (White and Voltmer 1965).

The voltage applied across the two electrodes creates an electric field that generates compressions and dilatations in the vicinity of the surface, giving rise to different elastic waves. Regarding the Rayleigh wave, emitted perpendicularly to the fingers of the comb-shaped electrodes, the transducer behaves like a series of ultrasonic sources. When the applied voltage is sinusoidal, the vibrations add up constructively only if the distance d is equal to half the Rayleigh wavelength λ_R. Indeed, any stress produced at time t by a pair of fingers travels at the velocity of the Rayleigh wave the distance $\lambda_R/2$ during the half-period $T/2$ (Figure 3.13(b)). At time $t + T/2$, this stress arrives under the neighboring pair of fingers, just when the voltage, which has changed sign, produces a stress with the same phase. The stress arising from the second pair of fingers adds constructively to the stress emitted by the first pair. The frequency $f_0 = V_R/2d$, which corresponds to this cumulative effect, is called the *synchronism frequency*. If the frequency deviates from this value, the interference between the elastic waves emitted by the different pairs of fingers is no longer totally constructive. The resulting signal is weaker; the bandwidth of such a transducer is narrower as it has more fingers.

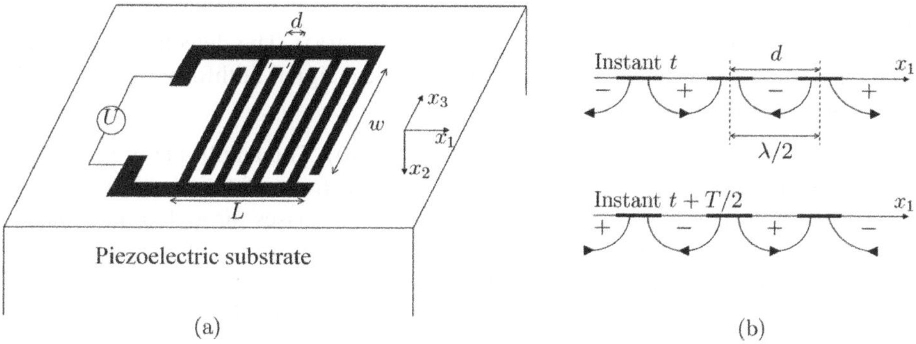

Figure 3.13. *a) Simple interdigital electrode transducer. b) The voltage applied across the two comb-shaped electrodes creates stresses near the surface of the solid*

The *frequency response* of a simple transducer with N pairs of fingers is deduced from its impulse response. A short pulse applied to the electrodes simultaneously excites the different parts of the transducer. Since the electric field is reversed at each interval between the fingers, the emitted wave train has a spatial period $2d$ and a time duration Θ equal to the active length of the transducer $L = 2Nd$, divided by the Rayleigh wave velocity V_R:

$$\Theta = \frac{2Nd}{V_R} = \frac{N}{f_0} \qquad \text{with} \qquad f_0 = \frac{V_R}{2d} \qquad [3.47]$$

By assimilating this impulse response of time duration Θ to a sinusoid of frequency f_0, the frequency response, equal to its Fourier transform, is a $(\sin X)/X$ function with:

$$X = \pi\Theta(f - f_0) = N\pi\frac{f - f_0}{f_0} \qquad [3.48]$$

The bandwidth at –3 dB ($X = \pm 0.885\pi/2$) is inversely proportional to the number of finger pairs N:

$$\frac{\Delta f}{f_0} = \frac{0.885}{N} \qquad [3.49]$$

In practice, to measure the impulse response of a transducer, it is necessary to transform the emitted elastic wave train into an electrical signal, using a receiver transducer of the same type. The electric field that accompanies the elastic wave induces a time-varying potential difference when it passes under the electrodes. The shape of the signal depends on the number of fingers in the receiver. Let us consider two typical cases:

a) receiver with only one finger pair (Figure 3.14(a)). The detection is localized along a line parallel to the wave fronts; the electrical signal faithfully reproduces the elastic signal as it passes under the receiver;

b) receiver identical to the emitter (Figure 3.14(b)). The amplitude of the voltage detected increases from time $\tau = l/V_R$, where the front of the elastic wave train reaches the first interdigital interval of the receiver; it passes through a maximum at time $\tau + \Theta$ when the different elements of the receiver are simultaneously excited and then decreases. The overall impulse response, with a triangular envelope and time duration 2Θ, is the autocorrelation function of the impulse response of one of the transducers. The frequency response of the line with two identical transducers is, consequently, a $[(\sin X)/X]^2$ function, whose relative bandwidth at –3 dB is $0.635/N$.

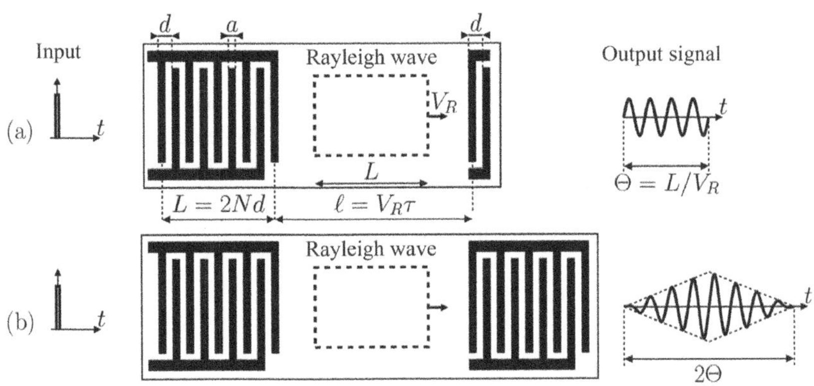

Figure 3.14. *Impulse response of a line with two transducers with comb-shaped electrodes deposited on a piezoelectric substrate. a) Receiver with two fingers and b) receiver identical to the emitter*

3.2.1.1. *Secondary effects*

This intuitive discussion assumes that the transformer operates ideally. In reality, parasitic effects occur.

3.2.1.1.1. Re-emission

The previous results are valid if the electromechanical coupling coefficient of the material is small or if the receiver is electrically short-circuited. On the other hand, the voltage induced at the terminals of the receiver transducer and the displacement of the electrical charges generate, through the inverse piezoelectric effect, a Rayleigh wave in each direction. This effect is significant if the electrical power extracted during the passage of the Rayleigh wave is a non-negligible fraction of the power of the incident wave. This re-emission due to the reciprocity of the piezoelectricity is analyzed by means of a 3-port circuit representing the transducer (section 3.2.3).

3.2.1.1.2. Bulk waves

The periodic structure also favors the emission of bulk waves with velocity V_B, in the directions $\pm\theta$, symmetrical with respect to the normal to the surface, and for which the conditions for constructive interference are satisfied: $2d \sin\theta = \lambda_B = V_B/f$. This effect only exists at frequencies higher than a cut-off value: $V_B/2d$. Since the velocity of bulk waves is generally greater than V_R, the frequency band corresponding to the generation of these waves is situated beyond the synchronism frequency of the Rayleigh wave transducer. These parasitic waves, partially reflected off the underside of the plate, are particularly annoying in the case of devices with a large bandwidth. These bulk waves can be partially eliminated by sanding the opposite face of the crystal or covering it with an absorbent.

3.2.1.1.3. Diffraction

In many applications, the width of the finger overlap must be varied. Consequently, the Rayleigh waves arising from sources whose width is of the order of the wavelength are diffracted (Chapter 1). The receiver then detects only part of the beam.

3.2.1.1.4. Phase velocity variations

Analyzing the beam perpendicularly to the propagation direction shows that the Rayleigh wave front emitted by the transducer is not parallel to the fingers. It is distorted when the finger length varies, especially if the electromechanical coupling coefficient of the substrate is large. This distortion arises from the non-homogeneity of the transducer: the velocity of the Rayleigh waves on the surface of a piezoelectric solid is affected by the presence of a metallic layer (section 3.1.3.2, Volume 1). When fingers of the comb-shaped electrodes are not of a constant length, the waves created by the various points on a line source travel paths of unequal metallized lengths and undergo different phase shifts.

3.2.1.1.5. Reflections

Variations in electrical and mechanical impedances at the surface, due to the metallic fingers, give rise to internal reflections. They are all the more important as the electromechanical coupling coefficient of the substrate is high and as the mechanical impedance of the finger material is very different from that of the substrate. They depend not only on the wave frequency but also on the transducer electrical load. Some structures have been designed to reduce these reflections, for example, the splitting of a finger into two parts. Indeed, the waves reflected on homologous edges of two ordinary neighboring fingers, of width $\lambda/4$, separated by an interval of $\lambda/4$, add constructively since the path difference causes a phase shift of 2π. These reflections are, in principle, suppressed when each finger is composed of two strips of width $\lambda/8$: the distance between the axes of two neighboring fingers remains $\lambda/2$, so there is destructive interference. However, these reflections, which have been the subject of many analyses and experiments (Datta and Hunsinger 1979, 1980), are now taken into account and exploited to produce unidirectional transducers (section 3.2.4), essential elements of low loss filters.

Let us now look at the discrete source model, or "delta" method, to calculate the response of the Rayleigh wave transducer.

3.2.1.2. *Discrete source model*

This model associates a delta function located, for example, at the center of each interval between the two comb-shaped electrodes (Tancrell and Holland 1971). The amplitude a_n of the source n is proportional to the overlap length w_n of the two fingers, with the sign + or – depending on the direction of the electric field (Figure 3.15).

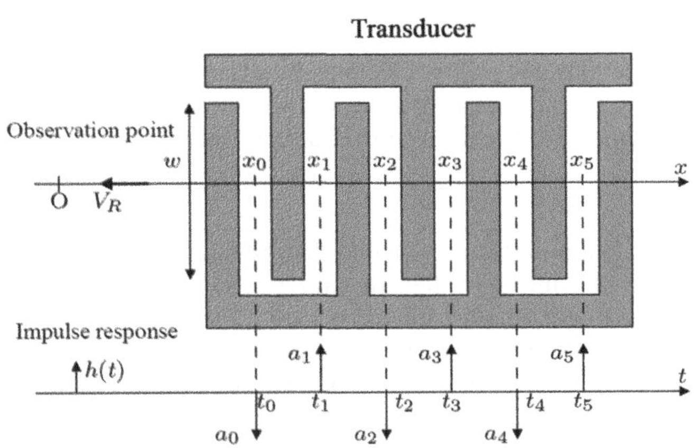

Figure 3.15. *Discrete source method. The transducer is assimilated to a series of sources located at the center (abscissa x_n) of the intervals, whose amplitudes a_n are proportional to the overlap lengths*

Under these assumptions, the impulse response of a transducer comprising N_s sources is a succession of Dirac functions:

$$h(t) = \sum_{n=0}^{N_s-1} s_n a_n \delta(t - t_n) \qquad [3.50]$$

The frequency response is equal to:

$$H(\omega) = \int_{-\infty}^{+\infty} h(t)e^{-i\omega t}\, dt = \sum_{n=0}^{N_s-1} s_n a_n e^{-i\omega t_n} \qquad [3.51]$$

where the sign s_n is equal to $(-1)^n$ when the direction of the electric field reverses at each interdigital interval. Let us apply this method of analysis to a simple transducer with N_s sources, such that:

$$s_n = e^{in\pi}, \qquad a_n = a_0 = \text{Const.}, \qquad t_n = t_0 + n\frac{d}{V_R} = t_0 + n\frac{\pi}{\omega_0} \qquad [3.52]$$

The frequency response is:

$$H(\omega) = a_0 e^{-i\omega t_0} \sum_{n=0}^{N_s-1} \left(e^{-2i\Delta\varphi}\right)^n \quad \text{by setting} \quad \Delta\varphi = \frac{\pi}{2}\frac{\omega - \omega_0}{\omega_0} \qquad [3.53]$$

The summation leads to:

$$H(\omega) = a_0 e^{-i\omega t_0} \frac{1 - e^{-2iN_s\Delta\varphi}}{1 - e^{-2i\Delta\varphi}} = a_0 \frac{\sin(N_s\Delta\varphi)}{\sin(\Delta\varphi)} e^{-i[\omega t_0 + (N_s - 1)\Delta\varphi]} \qquad [3.54]$$

Near the frequencies $f_m = (2m + 1)f_0$, which cancel the numerator and denominator:

$$\Delta\varphi = \frac{\pi}{2}\frac{f - f_m}{f_0} + m\pi \quad \text{and by setting} \quad X_m = N_s\frac{\pi}{2}\frac{f - f_m}{f_0} \qquad [3.55]$$

the frequency response varies as $(\sin X_m)/X_m$. In this discrete source model, the transducer response consists of bands of the same height, centered on the odd harmonic frequencies of the synchronism frequency f_0 (Figure 3.16). Their width at –3 dB ($X_m = \pm 0.885\pi/2$) is identical to [3.49] when the number of sources $N_s = 2N - 1$ is large. The relative frequency difference between two consecutive zeros is equal to $2/N_s$. The periodicity of the spectrum comes from the sampling (Shannon's theorem). The even harmonic bands disappear because the sampling function consists of two Dirac pulses with opposite signs and separated by $T_0/2$.

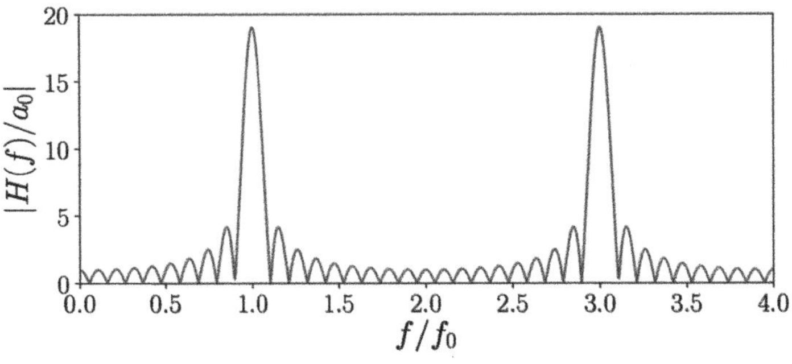

Figure 3.16. *Amplitude of the frequency response of a transducer with $N = 10$ pairs of fingers, calculated using the discrete source method*

The simple delta method only explains the shape of the frequency response of the transducer. In order to calculate the response amplitude and the radiation impedance of the transducer, it is necessary to consider the energy, with the help of other physical characteristics of the material.

3.2.2. *Impulse response model*

This method, based on energy considerations, offers the advantage to predict the absolute amplitude and thus the power of the transmitted Rayleigh waves. We assume that the transducer is uniform along x_1 and that the finger overlap width w is constant (Figure 3.14(a)). In order to calculate the response to an electrical excitation $U(t)$, let us introduce a scalar quantity $a(t)$ such that its square represents the elastic power density, per unit width of the Rayleigh wave beam, emitted in both directions $+x_1$ and $-x_1$ (Figure 3.17(a)). The square, $a^2(t)$, is expressed in terms of the Poynting vector component J_1 and of the average power density P_R transported by:

$$a^2(t) = \int_{-\infty}^{+\infty} J_1(x_2, t)\, \mathrm{d}x_2 \qquad \text{and} \qquad P_R = \langle a^2(t) \rangle \qquad [3.56]$$

In the harmonic case, a is proportional to the mechanical displacement or electric potential amplitude of the Rayleigh wave. In the impulse regime, $a(t)$ represents the instantaneous amplitude (Hartmann *et al.* 1973).

According to the relation [1.46] of Volume 1, and neglecting the emitted bulk waves, the application of Poynting theorem to a closed volume V containing the transducer, of width w, implies:

$$P_s(t) = \frac{\mathrm{d}E}{\mathrm{d}t} + 2\int_{-w/2}^{+w/2}\int_{-\infty}^{+\infty} J_1(x_2,t)\,\mathrm{d}x_2\,\mathrm{d}x_3 = \frac{\mathrm{d}E}{\mathrm{d}t} + 2wa^2(t) \qquad [3.57]$$

where P_s and E are, respectively, the power supplied by the electrical sources and the total energy stored within the volume V.

3.2.2.1. *Time responses*

Let us return to the reasoning and notations used in section 3.1.3.3 (Volume 1) by assuming that a continuous voltage of 1 volt was applied across the two electrodes of the transducer. These constitute a capacitance $C_T = NC_1w$ in which is stored the electrical energy:

$$E_{el} = \frac{1}{2}NC_1w \qquad \text{where} \qquad C_1 = \varepsilon_p + \varepsilon_0 \qquad [3.58]$$

is the capacitance per pair of fingers and per unit length. If, at time $t = 0$, the electrodes are short-circuited (Figure 3.17(b)), the response to this unit voltage step, $U(t) = 1 - H(t)$ is, by hypothesis, a sine function of frequency $f_0 = V_R/2d$ and time duration $\Theta = N/f_0$:

$$\begin{cases} a_u(t) = -a_0 \sin(2\pi f_0 t) & \text{for} \quad 0 < t < \Theta \\ a_u(t) = 0 & \text{for other values of } t \end{cases} \qquad [3.59]$$

a cosine function is not appropriate because the energy must remain continuous at the time origin.

Upon integrating relation [3.57], since $P_s = 0$ for $t > 0$, the acoustic energy generated is:

$$E_{ac} = E(0) - E(\infty) = 2w\int_0^{\infty} a_u^2(t)\,\mathrm{d}t = \frac{Nwa_0^2}{f_0} \qquad [3.60]$$

According to the definition of the electromechanical coupling coefficient K_R of the Rayleigh wave: $E_{ac} = K_R^2 E_{el}$, the amplitude of the unitary step response is equal to:

$$a_0 = K_R\sqrt{f_0 C_1/2} \qquad [3.61]$$

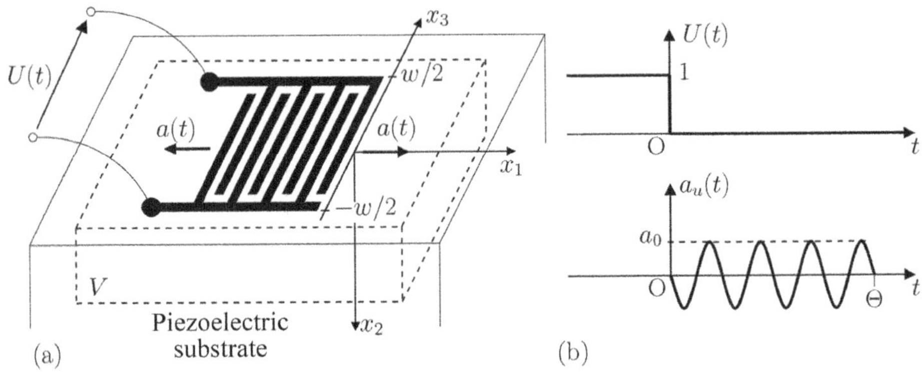

Figure 3.17. *a) Interdigital electrode transducer. b) Response to a unit voltage step*

The *impulse response* $[U(t) = \delta(t)]$ is the time-derivative of the step response:

$$\begin{aligned} h(t) &= -\frac{da_u}{dt} = \pi\sqrt{2}K_R\sqrt{C_1 f_0^3}\cos(2\pi f_0 t) \\ &= 2\pi a_0 f_0 \cos(2\pi f_0 t) \qquad \text{for} \qquad 0 < t < N/f_0 \end{aligned} \qquad [3.62]$$

The *frequency response*, that is, the Fourier transform of $h(t)$, varies as a $(\sin X)/X$ function (section 3.2.1).

If the finger overlap length w and the interdigital distance d vary with the abscissa x_1, that is, as a function of time $t = x_1/V_R$, the impulse response becomes:

$$h(t) = \pi\sqrt{2}K_R\sqrt{C_1 f_0^3}\frac{w(t)}{w_0}\cos\varphi(t) \text{ with } \varphi(t) = 2\pi\int_0^t f(t')\,dt' \qquad [3.63]$$

Conversely, an impulse response of the form $e(t)\cos\varphi(t)$ – and, thus, an imposed frequency response – is obtained by choosing the geometry of the transducer. The fingers are arranged on either side of lines of abscissa $x_n = V_R t_n$, with times t_n being defined by $\varphi(t_n) = n\pi$. The overlap length of two adjacent fingers is proportional to $e(t_n)[f(t_n)]^{-3/2}$.

As seen in the previous section, the detection of this elastic wave train through the inverse piezoelectric effect provides an output voltage similar to $h(t)$ when the receiver transducer is made up of two fingers whose length is at least equal to the maximum value of $w(t)$ (Figure 3.18).

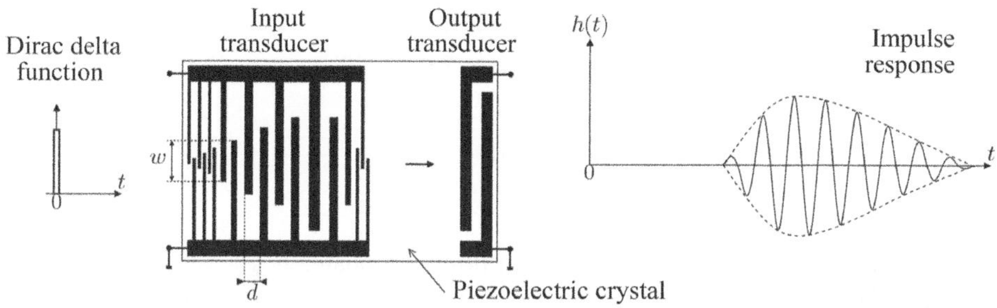

Figure 3.18. *Impulse response of a transducer: the variation in the finger overlap length determines the form of the envelope; the interdigital interval defines the instantaneous frequency*

Through convolution, the impulse response gives the response to any excitation. For example, if a sinusoidal voltage at the synchronism frequency and of unit amplitude: $U(t) = \cos(2\pi f_0 t)$ is applied to the transducer, the response is:

$$a_s(t) = U(t) \otimes h(t) = 2\pi a_0 f_0 \int_0^{\Theta} \cos\left[2\pi f_0(t-\tau)\right] \cos(2\pi f_0 \tau)\,\mathrm{d}\tau \qquad [3.64]$$

that is:

$$a_s(t) = \pi a_0 f_0 \int_0^{\Theta} \left\{\cos(2\pi f_0 t) + \cos\left[2\pi f_0(t-2\tau)\right]\right\} \mathrm{d}\tau \quad \text{with} \quad \Theta = \frac{N}{f_0} \qquad [3.65]$$

The second integral, which involves a multiple of the period T_0, is zero, so that:

$$a_s(t) = \pi N a_0 \cos(2\pi f_0 t) \qquad [3.66]$$

Compared with the unit step response [3.59], the synchronism effect multiplies the transducer response by $N\pi$. This effect is reduced when the frequency shifts away from f_0.

On the free surface, the amplitude of the electric potential accompanying the Rayleigh wave is related to the average power density transported by the wave through the formula [3.113] in Volume 1:

$$P_R = \langle a_s^2(t) \rangle = \frac{1}{2}\left(\pi N a_0 |U|\right)^2 = \frac{\pi^2}{4} N^2 K_R^2 f_0 C_1 |U|^2 \qquad [3.67]$$

where $|U|$ is the amplitude of the applied sinusoidal voltage. Given the expression for the capacitance per unit length $C_1 = \varepsilon_p + \varepsilon_0$, we obtain:

$$|\Phi_0| = \sqrt{\frac{2K_R^2}{\varepsilon_p + \varepsilon_0}\frac{P_R}{\omega_0}} = \frac{\sqrt{\pi}}{2}K_R^2 N|U| \qquad [3.68]$$

ORDER OF MAGNITUDE.– In the case of lithium niobate: $K_R^2 = 5\%$, with $N = 20$, the amplitude of the potential at the surface of the crystal is close to the amplitude of the voltage U applied to the transducer.

If the receiver is composed of many fingers of equal overlap length, greater than that of the largest fingers of the transmitter, the impulse response of the line formed by the two transducers is the convolution of the impulse response of the emitter with that of the receiver. As already seen in section 3.2.1, the envelope of the impulse response of a transmission line with two identical simple transducers (w and $d =$ Const) is a triangle of duration 2Θ. The response of a line consisting of two transducers, whose finger lengths vary, is more difficult to predict. An approximate solution is to mentally divide the line into parallel strips, to analyze the responses of each strip, and then to sum these responses. There is another method for weighting the response of a transducer: it consists of locally varying not the length but the density of the fingers by removing some of them. The amplitude of the waves emitted by a group is proportional to the number of the fingers.

3.2.2.2. *Radiation conductance*

The electrical impedance of an interdigital electrode transducer is made up of the static capacitance C_T and of a radiation impedance that expresses the emission of surface waves. The radiation conductance, which varies with frequency, is necessarily accompanied by an imaginary part $B_a(f)$, called the motional susceptance (Figure 3.19).

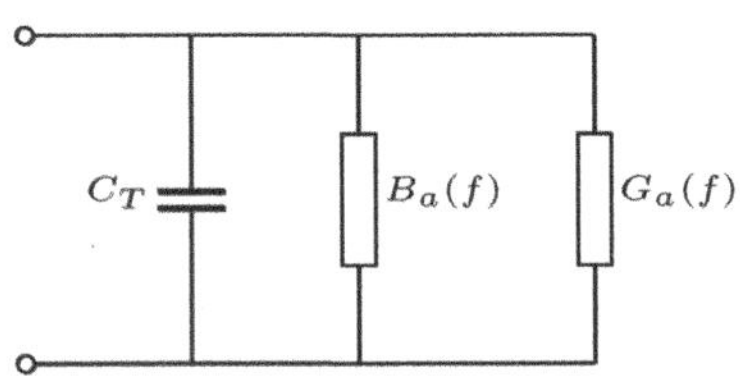

Figure 3.19. *Electrical admittance equivalent to a Rayleigh wave transducer:* $Y(f) = i2\pi C_T f + iB_a(f) + G_a(f)$

The radiation conductance $G_a(f)$ expresses that the mechanical energy generated by the transducer is equal to the electrical energy dissipated:

$$E_{ac} = 2w \int_0^{\infty} a^2(t)\,\mathrm{d}t = \int_{-\infty}^{\infty} G_a(f) \left|\hat{U}(f)\right|^2 \mathrm{d}f \tag{3.69}$$

$\hat{U}(f)$ is the Fourier transform of the applied voltage $U(t)$, so that $\hat{U}(f) = 1$, when $U(t) = \delta(t)$. Given the Parseval relation between the impulse response of the transducer, $h(t)$ and its frequency response $H(f)$:

$$\int_0^{\infty} h^2(t)\,\mathrm{d}t = \int_{-\infty}^{\infty} |H(f)|^2\,\mathrm{d}f \tag{3.70}$$

we obtain:

$$G_a(f) = 2w|H(f)|^2 \tag{3.71}$$

For the simple transducer whose impulse response is given by [3.63], the frequency response is formed of two $(\sin X)/X$ functions:

$$H(f) = \frac{\pi}{\sqrt{2}} N K_R \sqrt{C_1 f_0} \left(\frac{\sin X_+}{X_+} e^{-iX_+} + \frac{\sin X_-}{X_-} e^{-iX_-} \right) \quad \text{with} \quad X_\pm = N\pi \frac{f \pm f_0}{f_0} \tag{3.72}$$

If the relative bandwidth at –3 dB: $\Delta f / f_0 = 0.885/N$ is sufficiently narrow, so that the part of the spectrum centered on f_0 does not overlap the part centered on $-f_0$, the radiation conductance is given by:

$$G_a(f) = \pi^2 N K_R^2 C_T f_0 \left(\frac{\sin X}{X} \right)^2 = G_0 \left(\frac{\sin X}{X} \right)^2 \tag{3.73}$$

The value G_0 at the center frequency is proportional to the width w, through the intermediary of the total capacitance of the transducer: $C_T = NC_1 w$.

ORDER OF MAGNITUDE.– A transducer on lithium niobate (Y-cut, propagation along Z), operating at 100 MHz, consists of $N = 40$ finger pairs of overlap width $w = 50\lambda$ (= 1.75 mm). Using the constants of Table 3.1 in Volume 1, the total capacitance of the transducer is $C_T = 29$ pF and its radiation conductance at the center frequency: $G_0 = 0.055\ \Omega^{-1}$.

For any causal system, the imaginary part of the admittance $B_a(f)$ is the Hilbert transform of the real part $G_a(f)$:

$$B_a(f) = \frac{1}{\pi}\text{PP}\int_{-\infty}^{+\infty} \frac{G_a(\nu)}{\nu - f}\,\mathrm{d}\nu = G_0\frac{\sin(2X) - 2X}{2X^2} \qquad [3.74]$$

where PP denotes the principal part of the integral. Taking into account the static capacitance, the admittance of the transducer is written as:

$$Y(f) = G_0\left(\frac{\sin X}{X}\right)^2 + iC_T\omega_0\left(\frac{f}{f_0} + \pi K_R^2 N\frac{\sin(2X) - 2X}{4X^2}\right) \qquad [3.75]$$

Figure 3.20 is an example of the variation of the admittance Y of a simple transducer.

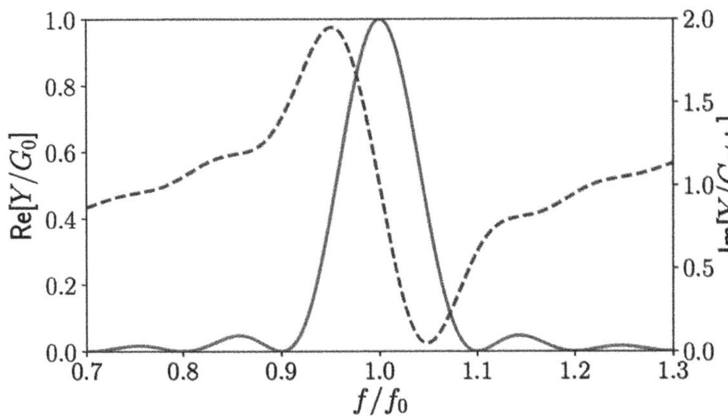

Figure 3.20. *Variation as a function of the normalized frequency, of the conductance (solid line) and the susceptance (dashed line) of a simple transducer with 10 finger pairs*

In order to ensure that the transducer is matched to the generator, the capacitance C_T is compensated by an inductance L, chosen so as to obtain the resonance at the synchronism frequency. In the case of the simple transducer, the bandwidth is therefore maximum for a value N_{opt} of the number of finger pairs, which depends on the electromechanical coupling coefficient. Indeed, according to formula [3.49], the "mechanical" bandwidth $(\Delta f)_m$ due to the array effect is inversely proportional to N. Taking into account the expression [3.73] for the conductance, the quality factor

Q and the bandwidth $(\Delta f)_e$ of the electric circuit tuned with the inductance L are equal to:

$$Q = \frac{C_T\omega_0}{G_0} = \frac{2}{\pi K_R^2 N} = \frac{f_0}{(\Delta f)_e} \qquad [3.76]$$

The bandwidth of the transducer is maximum when $(\Delta f)_m = (\Delta f)_e$:

$$\frac{0.885}{N} = \frac{\pi K_R^2 N}{2}, \quad \text{that is,} \quad N_{opt} = \frac{0,75}{K_R} \quad \text{and} \quad \frac{\Delta f}{f_0} = 1.18K_R \qquad [3.77]$$

ORDER OF MAGNITUDE.– For $LiNbO_3$, with $K_R^2 = 4.8\%$: $N_{opt} = 3.5$ and $\Delta f/f_0 = 26\%$.

The discrete source method has provided the shape of the frequency response, and the impulse response model has given the amplitude of this response and the radiation impedance. None of these models takes into account the reflections and re-emission within the transducer. In the following section, these phenomena are analyzed globally by using the scattering matrix.

3.2.3. *Three-port circuit (hexapole): scattering matrix*

The transducer with comb-shaped electrodes is, *a priori*, a device with two acoustic ports and one electrical port (hexapole). In the linear regime, its operation can be described, as for a bulk wave transducer (section 3.1.2.1), by a (3×3) matrix whose nature depends on the choice of variables. The quantities at the acoustic ports are, naturally, waves. If the same point of view is adopted for the electrical port, the scattering matrix $\boldsymbol{S}$, also called the distribution matrix, is homogeneous.

Let a_K be the complex amplitude of a wave, such that $a_K a_K^*$ represents a power in the harmonic case. This wave, entering by the port K of the three-port circuit in Figure 3.21, gives rise to a wave of amplitude b_J at the output J, such that: $b_J = s_{JK}a_K$. The nine dimensionless coefficients s_{JK} are *a priori* complex, and they constitute a matrix that expresses the three output quantities in terms of the three input quantities. The reciprocity theorem (the possibility of inverting the input and output quantities) reduces the number of independent coefficients to 6, the matrix $\boldsymbol{S}$ is symmetrical, $s_{JK} = s_{KJ}$:

$$\begin{pmatrix} b_1 \\ b_2 \\ b_3 \end{pmatrix} = \begin{pmatrix} s_{11} & s_{12} & s_{13} \\ s_{12} & s_{22} & s_{23} \\ s_{13} & s_{23} & s_{33} \end{pmatrix} \begin{pmatrix} a_1 \\ a_2 \\ a_3 \end{pmatrix} \quad \text{with} \quad s_{JK} = \sigma_{JK} e^{i\varphi_{JK}} \qquad [3.78]$$

s_{11}, s_{22}, s_{33} are the reflection coefficients, while s_{JK} $(J \neq K)$ are the transmission coefficients between the output J and the input K; σ_{JK}^2 is equal to the ratio of the output power at port J and the input power at port K; since the system is passive: $\sigma_{JK} \leq 1$.

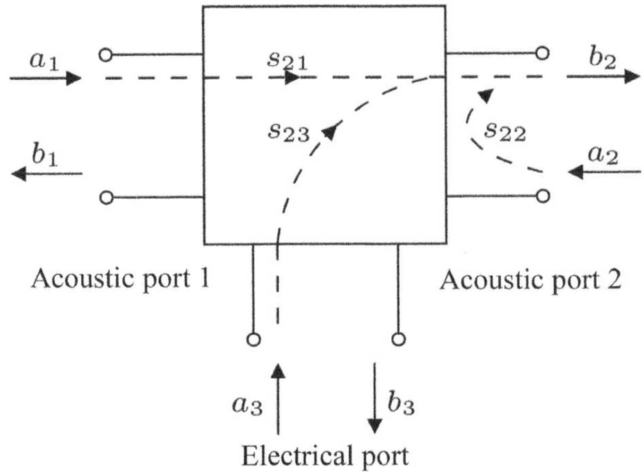

Figure 3.21. *The interdigital transducer is a device with one electrical port and two acoustic ports. It is characterized by the scattering matrix **S**. The connections in dashed lines symbolize the generation of wave b_2 by the three incoming waves a_1, a_2, a_3*

In the case where electrical losses (Joule effect) and acoustic losses (attenuation, generation of bulk waves) are negligible, the law of energy conservation imposes that the matrix be a unit matrix:

$$b_J b_J^\star = (s_{JI} a_I)(s_{JK} a_K)^\star = a_K a_K^\star \qquad \longrightarrow \qquad s_{JI} s_{JK}^\star = \delta_{IK} \tag{3.79}$$

that is:

$$s_{1I} s_{1K}^\star + s_{2I} s_{2K}^\star + s_{3I} s_{3K}^\star = \delta_{IK} \qquad \text{with} \qquad I, K = 1, 2, 3 \tag{3.80}$$

The expansion of this equation provides nine relations, as the last three, including a real part and an imaginary part, are double:

$$\sigma_{1K}^2 + \sigma_{2K}^2 + \sigma_{3K}^2 = 1 \qquad \text{with} \qquad K = 1, 2, 3 \tag{3.81}$$

and:

$$\begin{cases} s_{11} s_{12}^\star + s_{12} s_{22}^\star + s_{13} s_{23}^\star = 0 \\ s_{11} s_{13}^\star + s_{12} s_{23}^\star + s_{13} s_{33}^\star = 0 \\ s_{12} s_{13}^\star + s_{22} s_{23}^\star + s_{23} s_{33}^\star = 0 \end{cases} \tag{3.82}$$

In the case of a symmetric or antisymmetric transducer, if the transducer median line is taken as the origin, the two acoustic ports are indistinguishable: $s_{22} = s_{11}$ and a purely electrical excitation ($a_K = \delta_{K3}$) generates two acoustic waves, b_1 and b_2, that are equal or opposite ($s_{23} = \pm s_{13}$). There are four independent coefficients: s_{11}, s_{12}, s_{13} and s_{33}, and there remain five independent relations among [3.81] and [3.82]. If we consider only phase differences as meaningful, it is then possible to express six of the eight coefficients, σ and φ, in terms of the other two. For example, by referring to the conversion factor $\eta = \sigma_{13}^2 = \sigma_{23}^2$ and the angle $\phi = \varphi_{11} - \varphi_{12} \pm \pi$, it is found that:

$$\begin{aligned} &\sigma_{33}^2 = 1 - 2\eta \qquad \text{hence} \qquad \eta \leq 0.5, \\ &2\sigma_{11}\sigma_{12}\cos\phi = \eta \qquad \text{and} \qquad \sigma_{11}^2 + \sigma_{12}^2 = 1 - \eta \end{aligned} \qquad [3.83]$$

From the above considerations, it follows that the efficiency of the interdigital receiver does not exceed 50%. This limit, reached when the electrical matching to the load impedance is perfect ($\sigma_{33} = 0$), is a consequence of the transducer bidirectionality, as illustrated in Figure 3.22. An electrical excitation of unit power generates two symmetric waves, each transporting a power equal to 1/2. Reciprocally, two incident waves that are symmetrical with respect to the transducer are coupled, without any loss, to a perfectly matched transducer (Figure 3.22(a)). A single incoming wave (from the left) is equivalent to the superimposition of a symmetric mode and an antisymmetric mode (Figure 3.22(b)). Only the symmetrical pair generates an electric voltage at the transducer terminals. The antisymmetric pair passes under the transducer without any modification. Consequently, the interdigital electrode receiver extracts at most half the power of the incident Rayleigh wave: a fraction (at most equal to 1/4) of this is re-emitted (Figure 3.22(c)).

The resolution of the two equations [3.83] leads to the following expressions:

$$\begin{cases} 2\sigma_{12}^2 = 1 - \eta + \sqrt{1 - 2\eta - \eta^2 \tan^2\phi} \\ 2\sigma_{11}^2 = 1 - \eta - \sqrt{1 - 2\eta - \eta^2 \tan^2\phi} \end{cases} \qquad \text{with} \quad |\phi| < \arccos\left(\frac{\eta}{1-\eta}\right) \qquad [3.84]$$

The reflection coefficient of the incident Rayleigh wave on acoustic port "1" is minimum for $\phi = 0$. For the maximal efficiency value ($\eta = 0.5$), the reflection and transmission coefficients are equal: $\sigma_{11}^2 = \sigma_{12}^2 = 0.25$. For a line with two transducers, it follows that an electrical signal applied to the emitter gives several signals at the output of the receiver. Indeed, an electrical signal of unit power generates a Rayleigh wave with power η toward the receiver, assumed to be identical to the emitter. The receiver converts part of this (η) into an electrical signal (of power η^2) and re-emits the fraction $\eta\sigma_{11}^2$ toward the emitter. The emitter, in turn, re-emits the fraction of this incident wave (that is, $\eta\sigma_{11}^4$), which provides, after conversion by the receiver, an electrical signal of power $\eta^2\sigma_{11}^4$ and so on. The primary signal is thus followed by a series of secondary signals that appear at times 3τ, 5τ, etc. The first, which traveled thrice over the emitter-receiver distance, is the most important. In the extreme case $\sigma_{11} = 0.5$, the level of this "*triple-transit echo*" is $10 \log \sigma_{11}^4$, that is, -12 dB below that of the main signal. This echo, delayed by 2τ, causes, in the frequency response, oscillations of frequency $1/2\tau$. Given the expression [3.84] for σ_{11} in terms of η, the only means to lower the level of the triple-transit signal is to decrease the conversion factor, for example, by mismatching the impedances and thereby increasing the insertion losses of the line.

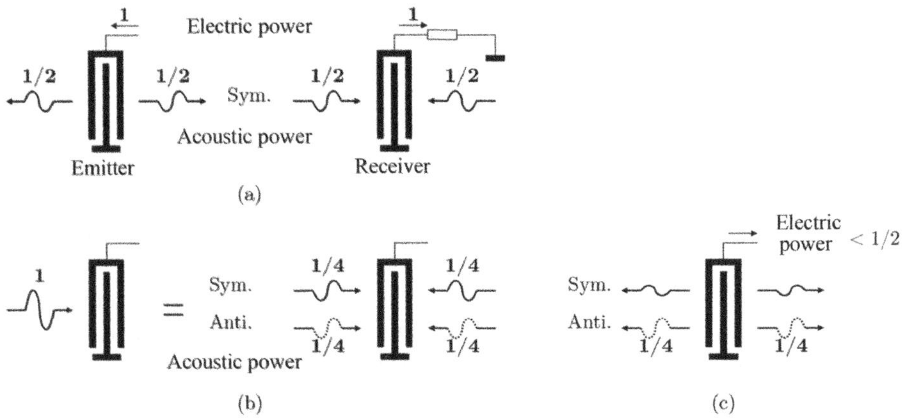

Figure 3.22. *Emitter and receiver operation of a symmetric interdigital transducer (according to Lewis (1983))*

As shown in Figure 3.22(a), two symmetrical waves are perfectly coupled to the electrical port of a symmetric transducer. The power conversion is total both in emission and reception. In Figure 3.22(b), the incident wave, of unit power arriving on the left of the receiver, is decomposed into symmetric and antisymmetric modes. Only the first mode is electrically coupled to the symmetric transducer, inducing an electrical signal whose power is at most equal to 1/2, depending on the transducer efficiency. If the electrical matching is not perfect, the symmetric mode is partially transmitted while the uncoupled antisymmetric mode is totally transmitted (Figure 3.22(c)). Then, transmitted modes interfere to give reflected and transmitted waves of different amplitudes. The same reasoning can be applied to an antisymmetric transducer.

Various means have been thought of for removing this parasitic echo. In a new generation of SAW devices (resonators with reflecting arrays, narrow band filters, rejection filters, etc.), a predominant role is played by reflection off the electrodes. These reflections are exploited to make the transducer unidirectional and reduce the losses in narrow-band filters (Hartmann *et al.* 1982). These reflections are locally taken into account in the coupled mode analysis.

3.2.4. *Coupled modes method*

When a Rayleigh wave enters a receiver transducer, it gives rise, while undergoing a periodic variation of velocity, to the following two effects:

– it induces a voltage across each pair of fingers, which generates two secondary waves. This re-emission, at the origin of the triple-transit echo in a transmission line, was discussed in section 3.2.3;

– it is partially reflected on the edges of each finger (section 3.1.1.1). These reflections lengthen the impulse response (infinitely, in the absence of losses).

As often happens, after having sought to eliminate the reflections, more or less successfully, the idea of exploiting them has emerged, for example, by compensating the piezoelectric re-emission in one direction by using internal reflections. The coupled modes analysis was thus developed (Chen and Haus 1985). It is first applied here to the case of a transducer with short-circuited electrodes, that is, to a simple reflective array, and then to an active transducer. The coupled mode method is based on the local analysis of the evolution of two coupled waves propagating in opposite directions. For this, the differential equation governing the evolution of the amplitude of each wave is established.

It must be recalled that in the absence of any transducer or array, the two waves, that are susceptible to propagate independently, in opposite directions on the substrate (with zero attenuation by hypothesis), obey the two uncoupled equations:

$$\frac{\mathrm{d}a}{\mathrm{d}x} = -ika \qquad \frac{\mathrm{d}b}{\mathrm{d}x} = ikb \qquad \text{with} \qquad k = k_R = \frac{\omega}{V_R} \tag{3.85}$$

The harmonic solutions take the form Ae^{-ikx} and Be^{ikx}, the factor $e^{i\omega t}$ being deliberately omitted.

The electrode array, assumed to be passive (the transducer is short-circuited), modifies the Rayleigh wave velocity by ΔV, and also the wave number, which becomes $k' = k_R + \Delta k$. On the other hand, the presence of the metallic strips gives rise to an exchange of energy between the two waves. The separation of the second-order propagation equation into two independent equations of the first order is no longer valid. However, if this energy exchange is small enough per period of the array, the unidirectional approximation can still be used, and the wave coupling can be considered as continuous. Thus, linear terms are introduced in equations [3.85], which translate into two coefficients, κ_{ab} and κ_{ba}:

$$\left\{\begin{array}{l}\dfrac{\mathrm{d}a}{\mathrm{d}x}=-ik'a-i\kappa_{ab}b\\[2ex]\dfrac{\mathrm{d}b}{\mathrm{d}x}=ik'b+i\kappa_{ba}b\end{array}\right.\qquad\text{with}\qquad k'=\frac{\omega}{V_R+\Delta V}\tag{3.86}$$

By hypothesis, the quantities a and b are such that the square of their modulus represents the power transported by the waves. Given the orthogonality of the two modes and their opposite propagation directions, the power flux is proportional to the difference $aa^\star - bb^\star$. If the energy between the two waves is exchanged without losses, the conservation of power imposes:

$$\frac{\mathrm{d}}{\mathrm{d}x}\left(aa^\star-bb^\star\right)=0\qquad\longrightarrow\qquad\kappa_{ba}=\kappa_{ab}^\star\tag{3.87}$$

Under the electrode array, of period p, the coefficient κ_{ab} is a periodic function of x, which can be expanded into a Fourier series:

$$\kappa_{ab}=\sum_{n=-\infty}^{n=+\infty}\kappa_{ab}^{(n)}e^{-inGx}\qquad\text{by setting}\qquad G=\frac{2\pi}{p}\tag{3.88}$$

Let us introduce two new functions $R(x)$ and $S(x)$:

$$a(x)=R(x)e^{-iGx/2}\qquad\text{and}\qquad b(x)=S(x)e^{iGx/2}\tag{3.89}$$

which reflect the slow variations of the amplitudes of the waves during their propagation. Equations [3.86] then become:

$$\left\{\begin{array}{l}\dfrac{\mathrm{d}R}{\mathrm{d}x}=-i\left(k'-\dfrac{G}{2}\right)R(x)-i\displaystyle\sum_{n=-\infty}^{n=+\infty}\kappa_{ab}^{(n)}S(x)e^{-i(n-1)Gx}\\[2ex]\dfrac{\mathrm{d}S}{\mathrm{d}x}=i\left(k'-\dfrac{G}{2}\right)S(x)+i\displaystyle\sum_{n=-\infty}^{n=+\infty}\left[\kappa_{ab}^{(n)}\right]^\star R(x)e^{i(n-1)Gx}\end{array}\right.\tag{3.90}$$

In first approximation, only the component $\kappa_{ab}^{(1)}=\kappa$, which provides a coupling term independent of x, is retained. Let us introduce the *detuning parameter*:

$$\delta=k'-\frac{G}{2}=\frac{\omega}{V_R+\Delta V}-\frac{\omega_0}{V_R}\approx\frac{\omega-\omega_0}{V_R}\qquad\text{if}\qquad\Delta V<<V_R\tag{3.91}$$

which is proportional to the difference between the angular frequency ω and the synchronism angular frequency of the array $\omega_0 = V_R G/2 = \pi V_R/p$. Thus, the coupled equations [3.90] are written as:

$$\frac{\mathrm{d}R}{\mathrm{d}x} = -i\delta R(x) - i\kappa S(x) \quad \text{and} \quad \frac{\mathrm{d}S}{\mathrm{d}x} = i\delta S(x) + i\kappa^\star R(x) \qquad [3.92]$$

3.2.4.1. *Passive array: reflection coefficient*

The previous analysis included the reflection by a passive array formed of metallic strips (short-circuited receiver transducer) or grooves (Figure 3.23). Indeed, any periodic modification of the surface impedance of the material, either by adjunction or by the removal of matter, generates reflections. The array of metallic strips must be short-circuited in order to prevent any piezoelectric re-emission. These reflectors play the role of mirrors in surface wave resonators.

The two coupled first order differential equations [3.92] are written in matrix form as:

$$\frac{\mathrm{d}\underline{X}}{\mathrm{d}x} = \boldsymbol{F}\underline{X}(x) \quad \text{by setting} \quad \underline{X} = \begin{pmatrix} R \\ S \end{pmatrix} \quad \text{and} \quad \boldsymbol{F} = \begin{pmatrix} -i\delta & -i\kappa \\ i\kappa^\star & i\delta \end{pmatrix} \qquad [3.93]$$

At a given frequency, the vector $\underline{X}$ describes, at each point of the array, the state of the surface waves propagating in opposite directions. The general solution is expressed in terms of the transition matrix $\boldsymbol{M}(x_1, x_0)$, which relates the amplitude vector of the waves at point x_0 to the amplitude vector at point x_1:

$$\underline{X}(x_1) = \boldsymbol{M}(x_1, x_0)\underline{X}(x_0) \qquad [3.94]$$

In the case of a *uniform array*, the parameters δ and κ only depend on the frequency and the matrix $\boldsymbol{F}$ is constant: the system is invariant according to x. The transition matrix $\boldsymbol{M}(x_1, x_0)$ is now only a function of the difference $x = x_1 - x_0$. It is equal to the exponential matrix $e^{\boldsymbol{F}x}$. Calculating the square of the matrix $\boldsymbol{F}$:

$$\boldsymbol{F}^2 = \begin{pmatrix} \kappa\kappa^\star - \delta^2 & 0 \\ 0 & \kappa\kappa^\star - \delta^2 \end{pmatrix} = \chi^2\mathbf{1} \text{ by setting } \chi^2 = \kappa\kappa^\star - \delta^2 = |\kappa|^2 - \delta^2 \qquad [3.95]$$

and the higher powers: $\boldsymbol{F}^3 = \chi^2\boldsymbol{F}$, $\boldsymbol{F}^4 = \chi^4\mathbf{1}$, $\boldsymbol{F}^5 = \chi^4\boldsymbol{F}$... leads to:

$$e^{\boldsymbol{F}x} = \cosh(\chi x)\mathbf{1} + \frac{\boldsymbol{F}}{\chi}\sinh(\chi x) \qquad [3.96]$$

Due to the conservation of transported power, the transition matrix is a unit matrix ($\det[\boldsymbol{M}(x)] = 1$). In the expression:

$$\boldsymbol{M} = \begin{pmatrix} \cosh(\chi x) - i\dfrac{\delta}{\chi}\sinh(\chi x) & -i\dfrac{\kappa}{\chi}\sinh(\chi x) \\ i\dfrac{\kappa^\star}{\chi}\sinh(\chi x) & \cosh(\chi x) + i\dfrac{\delta}{\chi}\sinh(\chi x) \end{pmatrix} \quad [3.97]$$

the parameter χ is real or purely imaginary depending on whether δ^2 is smaller or larger than $|\kappa|^2$.

In an array of length L (Figure 3.23), and at its two ends, $x_1 = L$ and $x_0 = 0$, the amplitudes $R(x)$ and $S(x)$ of the two coupled waves $a(x)$ and $b(x)$ propagating in opposite directions are given by [3.94] with $\boldsymbol{M}(L,0) = \boldsymbol{M}(L)$:

$$\begin{cases} R(L) = M_{11}(L)R(0) + M_{12}(L)S(0) \\ S(L) = M_{21}(L)R(0) + M_{22}(L)S(0) \end{cases} \quad [3.98]$$

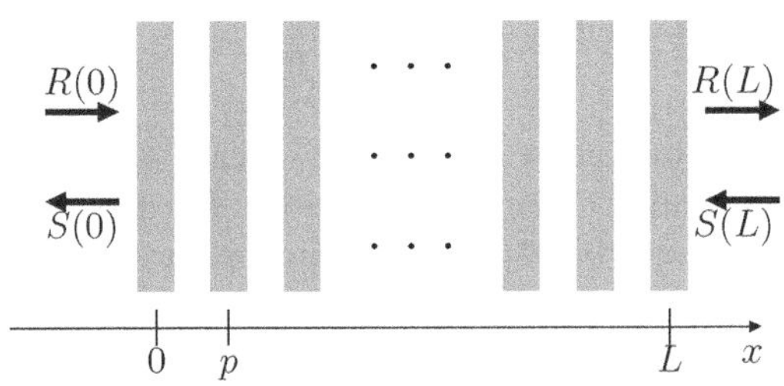

Figure 3.23. *Array of strips or grooves periodically modifying the surface impedance of a substrate*

In order to determine the *reflection coefficient* $r_0 = S(0)/R(0)$ of an incident wave of amplitude $R(0)$, arriving at the input $x = 0$, it is sufficient to write that at the other end of the array $S(L) = 0$:

$$r_0 = -\frac{M_{21}(L)}{M_{22}(L)} = \frac{-i\kappa^\star \sinh(\chi L)}{\chi\cosh(\chi L) + i\delta\sinh(\chi L)} \quad \text{with} \quad \chi^2 = |\kappa|^2 - \delta^2 \quad [3.99]$$

The modulus of r_0 passes through a maximum at the synchronism frequency of the array $f_0 = V_R/2p$, such that:

$$\delta = 0 \quad \longrightarrow \quad \chi = |\kappa| \quad \text{and} \quad |r_0(\delta = 0)| = \tanh(|\kappa|L) \quad [3.100]$$

A high reflectivity is achieved when $|\kappa|L > 1$, for example, $r_0 = 99.5\%$ for $|\kappa|L = 3$. An array of sufficient length acts as a mirror with high reflectivity in the vicinity of the Bragg frequency. When $|\delta| > |\kappa|$, $\chi = i\gamma$ is purely imaginary and the hyperbolic functions in [3.99] are replaced by trigonometric functions.

Figure 3.24 shows the variations in the reflection coefficient versus the product $L\delta$, which represents the deviation with respect to the frequency f_0 at which the Bragg condition is satisfied:

$$L\delta = \frac{2\pi}{V_R}(f - f_0)L = N\pi\frac{f - f_0}{f_0} \qquad [3.101]$$

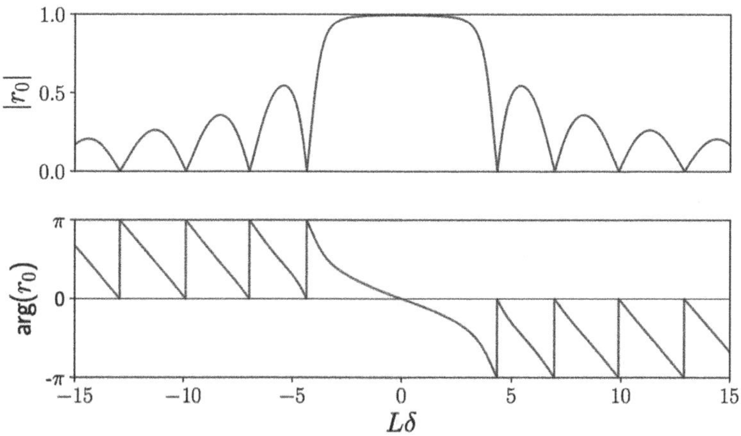

Figure 3.24. *Reflection coefficient r_0 of a periodic array calculated using the coupled modes theory ($\kappa L = -3i$). Amplitude and phase versus the dimensionless parameter $L\delta$*

The amplitude curve includes a central plateau and side lobes, separated by zeros at $\gamma L = \pm n\pi$. The bandwidth, defined by the gap between the two zeros flanking the plateau ($n = 1$), is given by:

$$L\delta = \sqrt{(|\kappa|L)^2 + \pi^2} \quad \longrightarrow \quad \frac{\Delta f}{f_0} = \frac{2}{N}\sqrt{1 + \left(\frac{|\kappa|L}{\pi}\right)^2} \qquad [3.102]$$

Figure 3.25 shows that in this frequency range, the amplitude of the mode R decreases within the array; for example, it can be established that at the center frequency:

$$\begin{cases} R(x) = \dfrac{\cosh\left[|\kappa|(L-x)\right]}{\cosh\left(|\kappa|L\right)} R(0) \\ S(x) = -i\dfrac{\kappa^\star}{|\kappa|}\dfrac{\sinh\left[|\kappa|(L-x)\right]}{\cosh\left(|\kappa|L\right)} R(0) \end{cases} \qquad [3.103]$$

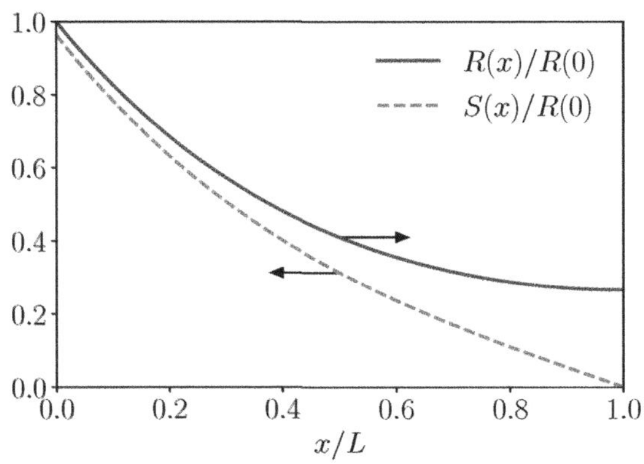

Figure 3.25. *Variation in the amplitude of the incident mode R and reflected mode S within a periodic array, when $\delta = 0$ and $|\kappa|L = 2$*

This decrease is due to the energy transfer, through the successive reflections, toward the mode S whose amplitude increases, taking into account its propagation in the opposite direction.

In the vicinity of the center frequency ($|\delta| << |\kappa|$ and $\chi \approx |\kappa|$), the phase of the reflection coefficient varies linearly with frequency (Figure 3.24), according to formula [3.99]:

$$\begin{aligned} \varphi &= -\frac{\pi}{2} - \arg\kappa - \arctan\left[\frac{\delta}{\chi}\tanh(\chi L)\right] \\ &\approx -\frac{\pi}{2} - \arg\kappa - \frac{\delta(\omega)}{|\kappa|}\tanh(|\kappa|L) \end{aligned} \qquad [3.104]$$

When $|\kappa|L > 1$, the reflection is accompanied by a constant group delay τ_g; the wave is delayed as if it were reflected from a plane located at an effective distance $L_{eff} = V_R\tau_g/2$ from the edge of the array, with:

$$\tau_g = -\frac{\mathrm{d}\varphi}{\mathrm{d}\omega} \approx \frac{1}{|\kappa|V_R} \qquad \text{and} \qquad L_{eff} = \frac{1}{2|\kappa|} \tag{3.105}$$

3.2.4.2. *Transducer*

If an electric voltage U is applied across the electrodes, the amplitudes R and S increase as the waves a and b pass between each pair of fingers. Since this growth is relatively small, it can be represented on the right-hand side of equations [3.92] by a term proportional to the voltage U:

$$\begin{cases} \dfrac{\mathrm{d}R}{\mathrm{d}x} = -i\delta R(x) - i\kappa S(x) + i\zeta U \\ \dfrac{\mathrm{d}S}{\mathrm{d}x} = i\delta S(x) + i\kappa^\star R(x) - i\xi U \end{cases} \tag{3.106}$$

The generation parameters ζ and ξ of the modes R and S are not independent. By hypothesis, the system is not dissipative; its evolution is reversible with respect to time. In the harmonic case, the change from t to $-t$ corresponds to the conjugation of the complex variables. In this operation, the propagation directions of the waves are reversed: $R^\star \rightarrow S$ and $S^\star \rightarrow R$. The solutions must satisfy equations [3.106], with $U^\star$ as the applied potential, which imposes $\xi = \zeta^\star$.

A third equation must be added to these two equations, giving the variation in current intensity ΔI (assumed to be small and continuous), from one pair of fingers to the next. This variation per unit length $(\Delta I/\Delta x)$ depends on the amplitude of the two waves propagating under the transducer and on the voltage U (Figure 3.26):

$$\frac{\mathrm{d}I}{\mathrm{d}x} = i\alpha R(x) + i\beta S(x) + iC\omega U \tag{3.107}$$

where Cp is the capacitance of a finger pair, p being the pitch of the strip array.

The new coefficients α and β are not independent of the generation parameters. Let us express, using the complex Poynting theorem (Volume 1, Appendix 3), that the electrical power supplied is entirely transmitted to the two modes. In a steady harmonic case (real $\omega \rightarrow \omega'' = 0$), and by taking the real part of relation ([A3.21], Volume 1), we get:

$$\frac{\mathrm{d}\langle P\rangle}{\mathrm{d}x} = \mathrm{Re}\left[\frac{1}{2}Uj^\star\right] \tag{3.108}$$

where:

$$\langle P \rangle = \text{Re}\,\boxed{P} = \frac{1}{2}\left(RR^{\star} - SS^{\star}\right) \quad \text{and} \quad j = \frac{\mathrm{d}I}{\mathrm{d}x} \qquad [3.109]$$

are, respectively, the average value of the transported power and the input current per unit length along x. After a few lines of calculations, the law of conservation of energy:

$$\frac{\mathrm{d}}{\mathrm{d}x}\left(RR^{\star} - SS^{\star}\right) = \text{Re}\left[U\frac{\mathrm{d}I^{\star}}{\mathrm{d}x}\right] = \frac{1}{2}\left(U\frac{\mathrm{d}I^{\star}}{\mathrm{d}x} + U^{\star}\frac{\mathrm{d}I}{\mathrm{d}x}\right) \qquad [3.110]$$

leads to the below equalities:

$$\alpha = -2\zeta^{\star} \qquad \text{and} \qquad \beta = -2\zeta \qquad [3.111]$$

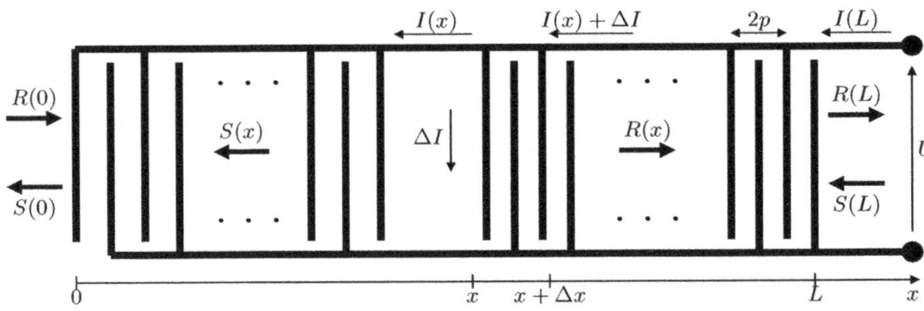

Figure 3.26. *Representation of an active transducer in the coupled mode model*

Finally, the variation in the intensity of the current supplied by the external source is given by:

$$\frac{\mathrm{d}I}{\mathrm{d}x} = -2i\zeta^{\star}R(x) - 2i\zeta S(x) + iC\omega U \qquad [3.112]$$

In the coupled mode model, the transducer is characterized by four parameters:

– the *capacitance* C per unit length, related to that of a finger pair $C_1 w$ by: $2pC = C_1 w$;

– the *detuning parameter*, given by [3.91]: $\delta = (\omega - \omega_0)/V_R$;

– the *reflection coefficient* κ, related to that r of a single finger for a short-circuited transducer: $r = i\kappa p$;

– the *generation factor* of a finger pair $2\zeta p$ deduced from a physical model, for example, based on the surface piezoelectric permittivity (section 3.1.3.2, Volume 1).

The amplitudes $R(x)$ and $S(x)$ of the two waves propagating in opposite directions are given by the coupled, first-order differential equations [3.106]. In matrix form, with $\xi = \zeta^\star$:

$$\frac{\mathrm{d}\underline{X}}{\mathrm{d}x} = \boldsymbol{F}\underline{X}\,(x) + \underline{G}\,(x)U \quad \text{with} \quad \underline{X} = \begin{pmatrix} R \\ S \end{pmatrix} \quad \text{and} \quad \underline{G} = \begin{pmatrix} i\zeta \\ -i\zeta^\star \end{pmatrix} \qquad [3.113]$$

they are analogous to the state equation [3.93], completed by the source term $\underline{G}\,U$. The general solution is expressed, using the transition matrix of the array $\boldsymbol{M}(x_1, x_0)$, through a convolution integral:

$$\underline{X}\,(x_1) = U \int_{x_0}^{x_1} \boldsymbol{M}(x_1, x)\underline{G}\,(x)\,\mathrm{d}x + \boldsymbol{M}(x_1, x_0)\underline{X}\,(x_0) \qquad [3.114]$$

3.2.4.2.1. Frequency responses of a uniform transducer

For a transducer of length L and at a given frequency, with $x_0 = L$ and $x_1 = 0$, the first line of system [3.114] leads to:

$$\begin{aligned} R(x_1 = 0, \omega) = {} & iU(\omega) \int_L^0 [M_{11}(0, x)\zeta(x) - M_{12}(0, x)\zeta^\star(x)]\,\mathrm{d}x \\ & + M_{11}(0, L)R(L, \omega) \end{aligned} \qquad [3.115]$$

The frequency response $H_R(\omega)$ of the transducer for the mode R is defined by the ratio $R(L, \omega)/U(\omega)$ in the absence of any incident wave at the end $x_1 = 0$ of the transducer, that is, when $R(0, \omega) = 0$:

$$H_R(\omega) = \frac{i}{M_{11}(0, L)} \int_0^L [M_{11}(0, x)\zeta(x) - M_{12}(0, x)\zeta^\star(x)]\,\mathrm{d}x \qquad [3.116]$$

Similarly, for the wave S, emitted in the opposite direction, the second line in equation [3.114] gives the ratio $H_S(\omega) = S(0, \omega)/U(\omega)$ with $x_1 = L$, $x_0 = 0$ and $S(L, \omega) = 0$:

$$H_S(\omega) = \frac{i}{M_{22}(L, 0)} \int_0^L [M_{22}(L, x)\zeta^\star(x) - M_{21}(L, x)\zeta(x)]\,\mathrm{d}x \qquad [3.117]$$

When the parameters δ and κ depend only on frequency, the matrix $\boldsymbol{F}$ (relation [3.93]) is constant and the transition matrix $\boldsymbol{M}(x_1 - x_0)$, which is now only a function of the difference $x = x_1 - x_0$, is given by equation [3.97]. This is true even if the generation factor ζ of the transducer depends on x, as happens when the finger overlap width varies. The calculation is simpler when ζ is constant (uniform transducer). Since $M_{11}(0, L) = M_{22}(L, 0) = M_{22}(L)$, relation [3.116] gives:

$$H_R(\omega) = \frac{1}{\chi M_{22}(L)} \int_0^L \left[(\kappa\zeta^\star - \delta\zeta) \sinh \chi x + i\chi\zeta \cosh \chi x\right] \mathrm{d}x \qquad [3.118]$$

that is, by integrating and replacing $M_{22}(L)$ by its expression:

$$H_R(\omega) = \frac{(\kappa\zeta^\star - \delta\zeta)(\cosh \chi L - 1) + i\chi\zeta \sinh \chi L}{\chi\,(\chi \cosh \chi L + i\delta \sinh \chi L)} \qquad [3.119]$$

For the wave emitted in the other direction, the same calculation leads to:

$$H_S(\omega) = \frac{(\kappa^\star\zeta - \delta\zeta^\star)(\cosh \chi L - 1) + i\chi\zeta^\star \sinh \chi L}{\chi\,(\chi \cosh \chi L + i\delta \sinh \chi L)} \qquad [3.120]$$

In the following section, the concepts of reflection center and generation center are explained and exploited to make the transducer unidirectional.

3.2.4.2.2. Generation center and reflection center

The expressions for the frequency responses show that the amplitudes $R(L)$ and $S(0)$ of waves emitted in opposite directions are not equal. The transducer is more efficient in one direction than the other. The directionality factor depends on the relative phase of the coefficients ζ and κ, that is, on the position of the *generation center* with respect to the position of the reflection center. To clarify these notions, let us consider a path Δx corresponding to a pair of fingers. The increase in the amplitude of the mode R is $\Delta R = G_1 U \Delta x$ with $\Delta x = 2p$. The increase in the amplitude of the mode S is $\Delta S = G_2 U \Delta x$, with $\Delta x = -2p$. G_1 and G_2 are the components of vector $\underline{G}$ in equation [3.113]. The two emitted waves: $\Delta R = i2p\zeta U$ and $\Delta S = i2p\zeta^\star U$ have the same amplitude; they are in phase if the parameter ζ is real. This point (line) from where the waves emitted in each direction are in phase is called the generation center. In the case of a uniform alternating transducer, the generation center of a finger pair is located in the middle of the active finger.

A *reflection center* is a point (line) from which waves are reflected with a phase shift equal to $\pm\pi/2$ (Wright 1989). The reflection center of a finger is located in its middle. In the case of a simple transducer, generation center and reflection center are merged. If the origin is chosen at the center of a finger, the parameters ζ and κ are real, so that $H_R(\omega) = H_S(\omega)$: the transducer is bidirectional. If the generation and reflection centers are separated, which can be achieved through various processes, the transducer may become unidirectional.

By taking as origin of the phases that of the generation factor: $\zeta = \zeta^\star$, the reflection coefficient is generally complex: $\kappa = \kappa' + i\kappa''$. At the center frequency, for which $\delta = 0$ and $\chi = |\kappa|$ is real, the frequency responses [3.119] and [3.120] are given by:

$$\begin{cases} H_R(\omega_0) = 2\zeta \dfrac{\sinh(\chi L/2)}{\chi^2 \cosh(\chi L)} D_R(\omega_0) \\ H_S(\omega_0) = 2\zeta \dfrac{\sinh(\chi L/2)}{\chi^2 \cosh(\chi L)} D_S(\omega_0) \end{cases} \qquad [3.121]$$

They are composed of a common factor and a directional term that takes into account the internal reflections:

$$\begin{cases} D_R(\omega_0) = \kappa \sinh\left(\chi \dfrac{L}{2}\right) + i\chi \cosh\left(\chi \dfrac{L}{2}\right) \\ D_S(\omega_0) = \kappa^\star \sinh\left(\chi \dfrac{L}{2}\right) + i\chi \cosh\left(\chi \dfrac{L}{2}\right) \end{cases} \qquad [3.122]$$

Since $\chi = |\kappa|$:

$$\begin{cases} |D_R(\omega_0)|^2 = |\kappa| \left[|\kappa| \cosh(\chi L) + \kappa'' \sinh(\chi L)\right] \\ |D_S(\omega_0)|^2 = |\kappa| \left[|\kappa| \cosh(\chi L) - \kappa'' \sinh(\chi L)\right] \end{cases} \qquad [3.123]$$

the *directionality factor* defined by the quantity:

$$F_D(\omega_0) = \frac{|D_R(\omega_0)|^2 - |D_S(\omega_0)|^2}{|D_R(\omega_0)|^2 + |D_S(\omega_0)|^2} = \frac{\kappa''}{|\kappa|} \tanh|\kappa| L \qquad [3.124]$$

is zero if $\kappa'' = 0$, that is, if the generation and reflection centers coincide: the transducer is bidirectional. Otherwise, the directionality factor is maximal when $\kappa' = 0$, that is, $\kappa'' = |\kappa|$. Its value is then equal to the reflection coefficient of the electrode array (formula [3.100]):

$$F_D(\omega_0) = \tanh|\kappa| L \qquad \longrightarrow \qquad \frac{|H_R(\omega_0)|}{|H_S(\omega_0)|} = e^{|\kappa| L} \qquad [3.125]$$

For example, if $|\kappa| L = 2$, the amplitude of the waves emitted in one direction is more than seven times larger than that of the waves emitted in the opposite direction. Internal reflections within the transducer favor the generation of Rayleigh waves in one direction, at the expense of that emitted in the other direction, provided that parameters

ζ and κ are $\pi/2$ out of phase. Given the round trip, this condition imposes that the reflection centers are shifted by $\lambda_0/8$ from the generation centers.

The coupled modes method has the advantage of providing analytical expressions for the main characteristics of a transducer and giving information on their evolution when a parameter varies. The model involves four parameters: the capacitance of a finger pair C_1, the detuning parameter δ, the reflection coefficient κp of a finger when the electrodes are short-circuited and the generation factor $2\zeta p$ of a finger pair. Their values must be determined by another analysis based on the constitutive equations, or through a series of experiments.

Furthermore, the structure of a transducer, a succession of electrodes and intervals, is discrete by nature. The model of coupled modes, which is based on the study of the variations, assumed to be continuous and slow, in amplitudes R and S of the two modes propagating in opposite directions, is thus an approximate approach. As the edge effects are not taken into account, this model applies well to long transducers with low internal reflections, that is, with a narrow bandwidth. Other models have been developed to take into account the discrete structure of a reflector or a transducer. The general procedure consists of first choosing an elementary cell and determining the elements of a transfer matrix, in order to calculate, through simple multiplication, a matrix specific to the complete reflector or transducer.

3.3. Generation by laser impact

Compared to the traditional methods (piezoelectric transducers, immersion techniques), photothermal generation offers several advantages: apart from the fact that it requires no mechanical contact, the position and shape of the source can be easily modified. Elastic waves can be generated in materials carried at high temperatures. This technique is presently oriented toward non-destructive testing, the measurement of elastic constants, acoustic emission, spectroscopy and acoustic microscopy. In most experiments, the solid is irradiated with short laser pulses. Bulk waves and surface waves are generated in this way. These elastic waves are detected either by conventional transducers (piezoelectric, capacitive, electromagnetic-acoustic) or by optical methods. Optical measurements, which will be studied in the next section, have the advantage of being carried out remotely, with a large bandwidth and without disturbing the acoustic field.

A pulsed laser beam, focused on the surface of a semi-infinite solid, into a point or along a line, acts as a point or line source that generates various elastic waves (Figure 1.17). In the core of the material, assumed to be isotropic, at time t after the impact of the laser pulse, the mechanical perturbation is localized on two circular arcs of radius $V_L t$ and $V_T t$ (Figure 1.18). On the free surface $x_2 = 0$, the perturbation due to the Rayleigh wave (R) appears at points of abscissa $x_1 = \pm V_R t$. An additional disturbance appears in the region $\theta > \theta_c = \arcsin(V_T/V_L)$, called the head wave (H), which originates from the fact that the longitudinal wave does not satisfy the boundary conditions at the free surface. However, it radiates energy toward the core of the solid; therefore, its amplitude decreases much more rapidly than that of the Rayleigh wave with the distance from the source.

The main parameters involved in the generation of elastic waves by a laser beam are those related to:

– *the laser pulse*:

- incident (absorbed) energy E (Q) (mJ);
- duration Δ (ns);
- area of the irradiated zone A (cm^2);
- density of absorbed power I (MW.cm^{-2});
- wavelength in vacuum Λ_0 (μm);

– *the material (isotropic)*:

- absorption coefficient η (%) : $Q = \eta E$;
- linear expansion coefficient α (K^{-1});
- mass density ρ (kg.m^{-3});
- heat capacity per unit mass C (J.kg^{-1}.K^{-1});
- thermal diffusivity D_{th} (m^2.s^{-1});
- Lamé constants λ, μ (GPa);
- bulk modulus K (GPa).

The most commonly used sources are the Nd:YAG laser, made of a crystal of yttrium aluminum garnet ($Y_3Al_5O_{12}$) doped with neodymium (Nd), and the carbon dioxide (CO_2) laser. The wavelengths of emitted radiations are, respectively, 1.064 μm and 10.6 μm, for pulse durations from 10 to 100 ns and pulse rates ranging from 1 to 100 shots per second.

Depending on the power density, the impact of a light pulse on the free surface of an opaque solid generates elastic waves through different mechanisms: those that bring about a modification in the surface state (ablation phenomenon) and those that do not modify it (radiation pressure, thermoelastic effect). When the light power density does not result in any vaporization of matter, the local expansion resulting from the heating (thermoelastic regime) is predominant, compared to the radiation pressure effect. This dilatation creates forces that are essentially parallel to the free surface

(Figure 3.27(a)). If the absorbed light intensity is sufficient ($I > 15$ MW/cm^2 for duralumin), the incident laser pulse causes a vaporization of matter. This ablation regime, which is accompanied by a transfer of momentum, creates forces that are essentially normal to the surface (Figure 3.27(b)).

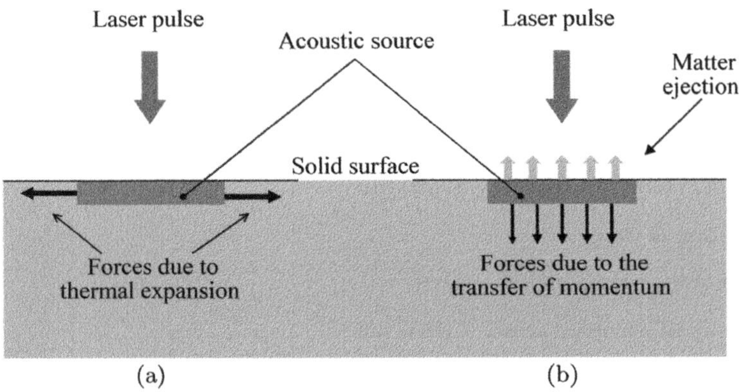

Figure 3.27. *Photothermal generation: a) thermoelastic regime; b) ablation regime*

3.3.1. *Thermoelastic regime*

When a radiation of medium intensity falls on a metallic surface, the electromagnetic field induces a conduction current near the surface. A part of the incident energy is absorbed and converted into heat by Joule effect, while the other part is reflected. Because of the screening effect of the conduction electrons, phenomena are localized very close to the surface of the metal, in the skin thickness. The laser pulse acts as a thermal energy source and generates a mechanical deformation. If the laser beam is focused inside the solid, the point-like thermoelastic source creates only spherical longitudinal waves. The presence of the surface is at the origin of a conversion into transverse waves. The combination of longitudinal and transverse polarizations generates Rayleigh waves, which propagate at the surface of the solid or Lamb waves, in the case of a plate.

3.3.1.1. *Temperature distribution*

The electromagnetic wave penetrates into the metal up to a depth equal to the *skin thickness* γ, which depends on the wavelength Λ_0, the electrical conductivity σ and the permeability μ_m of the metal. The absorption coefficient η is given by:

$$\eta = \frac{4\pi\gamma}{\Lambda_0} \quad \text{with} \quad \gamma = \sqrt{\frac{\Lambda_0}{\pi\sigma c_0 \mu_m}} \quad \text{and} \quad c_0 = 3 \times 10^8 \text{ m/s} \qquad [3.126]$$

ORDER OF MAGNITUDE.– For duralumin ($\sigma = 4 \times 10^7\ \Omega^{-1}\text{m}^{-1}$, $\mu_m = \mu_0 = 4\pi \times 10^{-7}$ H.m^{-1}) and a radiation coming from a YAG laser ($\Lambda_0 = 1.06\ \mu$m), the penetration depth is of the order of 5 nm and the fraction of incident energy absorbed is of the order of 6%–7%.

The depth of the thermoelastic source is determined by the thermal diffusion process. The temperature rise T is governed by the heat equation (Carslaw and Jaeger 1959):

$$\frac{\partial T}{\partial t} - D_{th}\Delta T = \frac{1}{\rho C} P_a(\underline{x}, t) \qquad [3.127]$$

where D_{th} is the thermal diffusivity and $P_a = \eta P$ is the power absorbed per unit volume.

With $q(t)$ denoting the time variation of the light power (normalized to unity) and by assuming a Gaussian spatial distribution of radius a for the laser beam intensity, the power absorbed per unit volume can be written as:

$$P_a = f(z)g(r)q(t) \quad \text{with} \quad f(z) = \frac{Q}{\gamma} e^{-z/\gamma} \quad \text{and} \quad g(r) = \frac{e^{-(r/a)^2}}{\pi a^2} \qquad [3.128]$$

Neglecting the thermal diffusion in the air, Ready has established the expression for the temperature variation in the metal (Ready 1971):

$$T - T_0 = \frac{Q}{\pi\rho C\sqrt{D_{th}}} \int_0^t \frac{q(t-t')}{\sqrt{\pi t'}(4D_{th}t' + a^2)} \times \exp\left(-\frac{z^2}{4D_{th}t'} - \frac{r^2}{4D_{th}t' + a^2}\right) \mathrm{d}t' \qquad [3.129]$$

If the material diffusivity D_{th} and the time duration Δ of the laser pulse are sufficiently small, so that $D_{th}\Delta << a^2$, the transverse diffusion is negligible: the radial distribution of temperature follows the profile of the absorbed light intensity. This temperature distribution is represented in Figure 3.28 in the case of duralumin, for a laser beam with diameter $2a = 1$ mm (absorbed energy $Q = 1$ mJ) and a pulse shape $q(t)$ similar to that in Figure 1.24, of time duration $\Delta = 2.4\tau = 24$ ns. The penetration of the thermal wave remains smaller than 10 μm, even 100 ns after the impact of the light pulse. During its duration Δ, the diffusion depth reaches about 2 μm, which is very small compared to the light beam diameter and the acoustic wavelength.

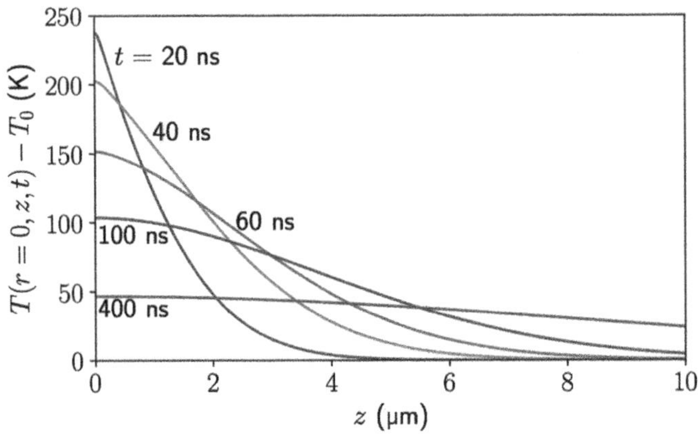

Figure 3.28. *Temperature distribution in duralumin versus depth at different times. Time duration of the incident laser pulse: $\Delta = 24$ ns, absorbed energy: $Q = 1$ mJ. For a color version of this figure, see www.iste.co.uk/royer/waves2.zip*

The thermoelastic source is localized below the surface, in a very thin disk of thickness d, of the micrometer order, and of area $A = \pi a^2$. The local heating disappears after a time of the order of 400 ns. Pulses arriving at a rate less than 1 MHz are independent.

3.3.1.2. *One-dimensional model*

In this simple model, the thermal diffusion over the duration (10–100 ns) of the laser pulse is neglected. In the case of a very short laser pulse of energy E, arriving on the surface at time $t = 0$, the power absorbed per unit volume can be written as:

$$P_a(t) = \frac{\eta E}{Ad}\delta(t) = \frac{Q}{V}\delta(t) \qquad [3.130]$$

where $\delta(t)$ is the Dirac delta function and A the cross-sectional area of the laser beam. Neglecting the diffusion term in the heat equation [3.127], the temperature rise $\Delta T = T - T_0$ of the heated volume $V = Ad$ varies as the step function $H(t)$:

$$\Delta T(t) = \frac{1}{\rho C}\int_0^t P_a(t')\,\mathrm{d}t' = \frac{Q}{\rho C V}H(t) \qquad [3.131]$$

This causes a dilatation ΔV of the volume V:

$$\Delta V = 3\alpha V \Delta T(t) = \frac{3\alpha}{\rho C}QH(t) \qquad [3.132]$$

proportional to the absorbed energy Q and to the *thermoelastic parameter* $(\alpha/\rho C)$; for duralumin:

$$\frac{\alpha}{\rho C} = 0.84 \times 10^{-11}\,\mathrm{m^3 J^{-1}} \qquad \text{and} \qquad \Delta V(\mu\mathrm{m}^3) = 2.5 \times 10^4 Q(\mathrm{mJ}) \qquad [3.133]$$

The temporal shape of the expansion is only slightly modified when thermal diffusion is taken into account; it remains very close to a step function, apart from a rise time of a few tens of nanoseconds.

The irradiated area is not free to expand radially, due to the rigidity of the surrounding material. The resulting mechanical stresses σ_{ij} can be expressed in terms of the strains ε_{kl} and the temperature rise $\Delta T = T - T_0$ by the relation:

$$\sigma_{ij} = c_{ijkl}\left(\varepsilon_{kl} - \alpha_{kl}\Delta T\right) = c_{ijkl}\varepsilon_{kl} - \Delta\sigma_{ij} \qquad [3.134]$$

where $\Delta\sigma_{ij} = c_{ijkl}\alpha_{kl}\Delta T$ represents the stresses exerted by the thermoelastic source on the surrounding material. In the case of an isotropic solid, taking into account the expressions for the stiffness tensor c_{ijkl} (equation [1.66], Volume 1) in terms of the Lamé constants λ and μ and the expression of the expansion tensor $\alpha_{kl} = \alpha\delta_{kl}$, we get:

$$\sigma_{ij} = \lambda\varepsilon_{kk}\delta_{ij} + 2\mu\varepsilon_{ij} - (3\lambda + 2\mu)\alpha\Delta T\delta_{ij} \qquad [3.135]$$

If the lateral dimensions of the source are large compared to the acoustic wavelength, the strain ε_{22}, along the normal x_2 to the irradiated surface, is preponderant. Since the normal stress:

$$\sigma_{22} = (\lambda + 2\mu)\varepsilon_{22} - (3\lambda + 2\mu)\alpha\Delta T \qquad [3.136]$$

cancels on the free surface $x_2 = 0$, the strain is given by:

$$\varepsilon_{22} = \frac{3\lambda + 2\mu}{\lambda + 2\mu}\alpha\Delta T = \frac{\partial u_2}{\partial x_2} \qquad [3.137]$$

In this one-dimensional model, the source of piston-type generates only longitudinal waves perpendicular to the surface, of amplitude $u_2 = \varepsilon_{22}d$, that is, given the relation [3.131]:

$$u_2 = \frac{3\lambda + 2\mu}{\lambda + 2\mu}\frac{\alpha}{\rho C}\frac{Q}{A}H(t) \qquad [3.138]$$

Let us note that this displacement does not depend on the diffusion depth d but on:

– a dimensionless coefficient $(3\lambda + 2\mu)/(\lambda + 2\mu)$, which is equal to 2, in the case of a material like duralumin, with Poisson's ratio $\nu \approx 1/3 \longrightarrow \lambda \approx 2\mu$;

– the thermoelastic parameter of the material: $\alpha/\rho C$;

– the energy density absorbed per unit area: Q/A.

ORDER OF MAGNITUDE.– With the value given in equation [3.133] and for an absorbed energy density Q/A = 10 mJ/cm^2, the displacement of matter is: $u_2 =$ 1.68 nm.

This model is valid as long as the lateral dimensions of the source are large with respect to the acoustic wavelength. In practice, given the time duration of the laser pulse (10 to 50 ns), this is a fraction of a millimeter. This condition leads to a large beam diameter ($2a \approx 1cm$), and consequently to power density and displacement amplitude that are too small to be optically detected.

3.3.1.3. *Point source model*

The radiation by a thermoelastic source of arbitrary dimensions can be calculated from the radiation of a point-like source. The thermoelastic expansion creates a sudden variation in the stresses $\Delta\sigma_{ij}$ in the heated zone. In seismology, Aki and Richards (1997) have established that the displacement generated by this type of "explosive" source is given by:

$$u_n(\underline{x}\,,t) = \int_V \Delta\sigma_{ij}(\underline{\xi}\,,t) \otimes \frac{\partial G_{ni}(\underline{x}\,,\underline{\xi}\,,t)}{\partial \xi_j}\,\mathrm{d}\underline{\xi} \qquad [3.139]$$

$G_{ni}(\underline{x}\,,\underline{\xi}\,,t)$ is the Green's function giving the nth displacement component at the observation point $\underline{x}$ and at time t, due to an impulsive force parallel to the x_i axis and applied at the source point $\underline{\xi}$ and at time $t = 0$ (Figure 3.29(a)). The symbol $\otimes$ denotes the convolution operation on the time variable. This expression simplifies if the source is assumed to be point-like and localized at the origin of the coordinate system ($\underline{\xi} = \underline{0}$):

$$u_n(\underline{x}\,,t) = M_{ij}(t) \otimes \left.\frac{\partial G_{ni}(\underline{x}\,,\underline{\xi}\,,t)}{\partial \xi_j}\right|_{\underline{\xi}=\underline{0}} \qquad [3.140]$$

where:

$$M_{ij}(t) = \int_V \Delta\sigma_{ij}(\underline{0}\,,t)\,\mathrm{d}\underline{\xi} = V\Delta\sigma_{ij}(\underline{0}\,,t) \qquad [3.141]$$

is the tensor of seismic moment modeling the point source.

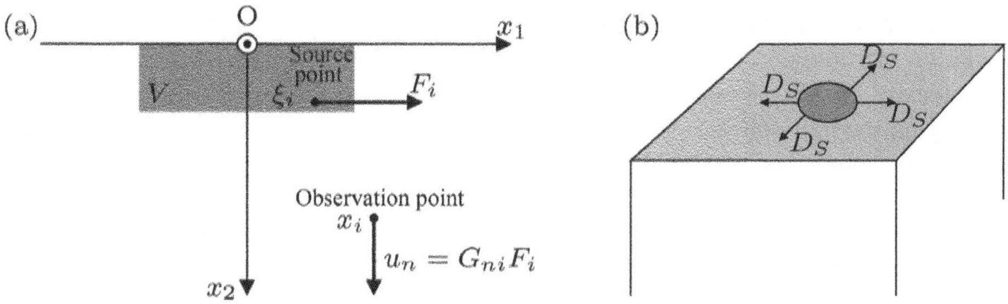

Figure 3.29. *a) The component F_i of the force applied at the source point $\underline{\xi}$, contained in the volume V, creates a displacement $u_n = G_{ni}F_i$ at the observation point $\underline{x}$. b) Dipoles of forces equivalent to a surface thermoelastic source*

For an isotropic solid, given expression [3.135], we get:

$$\Delta\sigma_{ij} = (3\lambda + 2\mu)\alpha\Delta T\delta_{ij} = K\frac{\Delta V}{V}\delta_{ij} \qquad \text{where} \qquad K = \lambda + \frac{2}{3}\mu \quad [3.142]$$

is the bulk modulus. The tensor of seismic moment reduces to a scalar:

$$M_{ij}(t) = K\Delta V(t)\delta_{ij} \quad [3.143]$$

which characterizes the intensity of the source and whose time variation reproduces the step-variation in the irradiated volume. Introducing this tensor into equation [3.140] reveals the vector $g_n^H(\underline{x}, \underline{0}, t)$, equal to the divergence, at the origin $\underline{\xi} = \underline{0}$, of the Green's tensor G_{ni}:

$$g_n^H(\underline{x}, \underline{0}, t) = H(t) \otimes \left.\frac{\partial G_{ni}}{\partial \xi_i}\right|_{\underline{\xi}=\underline{0}} \quad [3.144]$$

corresponding to a step-shaped time variation, rather than a delta function. The components of the mechanical displacement are given by:

$$u_n(\underline{x}, t) = K\Delta V g_n^H(\underline{x}, \underline{0}, t) \qquad \text{with} \qquad K\Delta V = (3\lambda + 2\mu)\frac{\alpha}{\rho C}Q \quad [3.145]$$

Its amplitude is proportional to the absorbed energy Q, that is, to the incident light energy in the thermoelastic regime. In the case of a time dependence $q(t)$ of the laser power, the displacement is obtained through convolution:

$$u_n(\underline{x}, t) = K\Delta V q(t) \otimes g_n^H(\underline{x}, \underline{0}, t) \quad [3.146]$$

This result can also be obtained by modeling the stresses due to the sudden variation in volume ΔV by a set of three orthogonal force dipoles (Achenbach 1987) of intensity $K\Delta V$. A thermoelastic source located on the free surface of a solid occupying the half-space $x_2 > 0$ only creates tangential forces: $G_{n2} = 0$. Therefore, the point source is modeled by two orthogonal and horizontal force dipoles (Figure 3.29(b)) and is equivalent to a center of expansion (Rose 1984). Nonetheless, since the source is on the surface, the intensity D_S of the two dipoles reflecting the expansion of matter is reduced (Royer 2001):

$$D_S = \frac{2\mu}{\lambda + 2\mu} K\Delta V = \frac{1-2\nu}{1-\nu} K\Delta V \qquad [3.147]$$

For a material with the Poisson's ratio $\nu = 1/3$, such as duralumin, D_S is half the intensity $K\Delta V$ of the three dipoles equivalent to the same thermoelastic source in the bulk of the material. This effect is due to the absence of matter in the upper half-space, which reduces the surface stiffness.

In order to determine the mechanical displacements generated in the solid, we must calculate the Green's function $\underline{g}^H = \text{div}\, \underline{\underline{G}}^H$. The generation of bulk waves will be studied first and then that of surface waves.

3.3.1.3.1. Bulk waves

Mechanical displacements generated inside the material at a distance r from the laser impact can be classified into two categories. In the near field, displacements are observed around the arrival times r/V_L and r/V_T, of the longitudinal and transverse wave fronts, respectively, and between these times and after the transverse wave front. In the far field, displacements are localized around longitudinal and transverse arrival times, and their amplitudes only depend on the direction.

Near field

Due to symmetry, the displacement along the x_2 axis, that is, at the epicenter, with respect to the source, is normal to the free surface. Figure 3.30 shows the normal component u_2, at three depths within a material with Poisson's ratio $\nu \approx 1/3$. The source is a laser pulse of Gaussian profile (diameter 2 mm), of temporal shape $q(t)$ and duration 12 ns (see Figure 1.24). The density of absorbed energy is equal to 10 mJ/cm^2. In Figure 3.30(a), the thermal diffusivity D_{th} and skin thickness γ are assumed to be zero, so that the thermoelastic source is localized on the surface. The first negative variation, propagating at velocity V_L, corresponds to a matter withdrawal at the epicenter, in the direction normal to the surface. This negative step is followed by a slow depression and a positive step, propagating at the transverse wave velocity $V_T \approx V_L/2$, corresponding to a shear strain. After this wave front, the sample tends to return to an equilibrium state different from the rest state because the source (in step) is still expanding. The slow variation between these two fronts is characteristic of the coupling between L and T waves in the near field. The mechanical displacement decreases rapidly with depth because the thermoelastic source is not very directional.

Figure 3.30(b) represents the displacement at the epicenter, calculated by taking into account the thermal diffusion ($D_{th} = 6 \times 10^{-5}$ m^2/s) and the optical penetration depth ($\gamma = 5$ nm). Due to these two phenomena, the thermoelastic source penetrates a few μm inside the solid and creates a dipole of forces normal to the surface, which generates a short, positive, longitudinal displacement. For these high-frequency spectral components ($f \approx 20$ MHz for $\tau = 5$ ns, see Figure 1.24), the source, whose diameter is about six times larger than the wavelength of the longitudinal wave, operates in the piston mode. Since its radiation is much more directional than that of the surface center of expansion, the peak amplitude decreases little with depth.

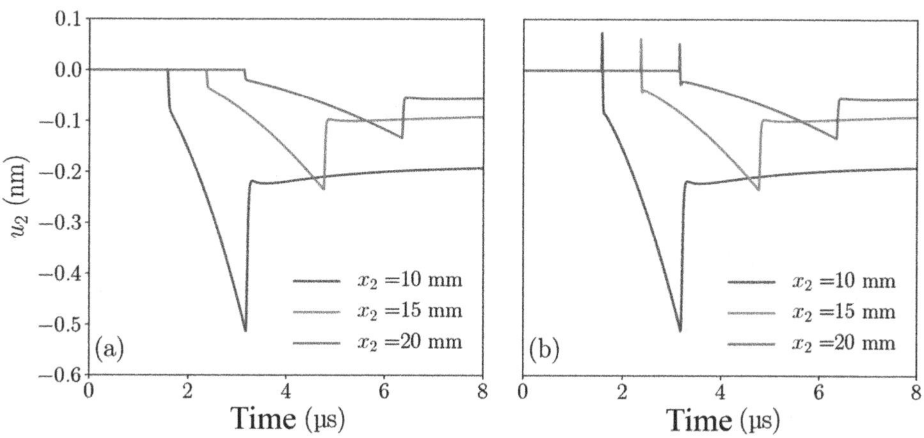

Figure 3.30. *Generation, in a duralumin block, by a laser pulse with incident energy 5 mJ and a half-height diameter of 2 mm. Mechanical displacement calculated at the epicenter at several depths x_2 = 10, 15 and 20 mm: a) without taking into account thermal diffusion and optical penetration, b) taking into account these phenomena. For a color version of this figure, see www.iste.co.uk/royer/waves2.zip*

Far field, directivity diagrams

In the case of a line source along the x_3 axis, the forces $\pm F_3$ of the dipoles cancel each other out. The thermoelastic source is equivalent to two lines of opposite forces of density per unit length $\pm F_1$ that are spaced apart by a distance a (Figure 3.31).

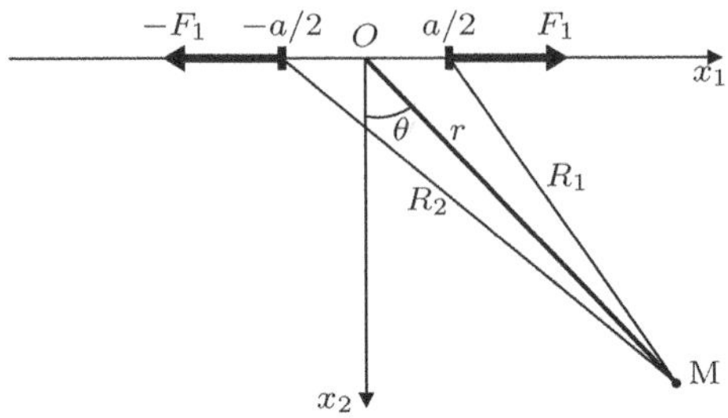

Figure 3.31. *Dipole of forces equivalent to a line thermoelastic source*

With M = L or T, depending on the type of wave, the mechanical displacement generated by the dipole of forces, with intensity per unity length $D_S = F_1 a$, at a distance r large in front of a, is equal to:

$$\begin{aligned}\underline{u}^D &= \underline{u}^F(\theta, t - s_M R_1) - \underline{u}^F(\theta, t - s_M R_2) \\ &\approx s_M(R_2 - R_1)\frac{\partial \underline{u}^F}{\partial t} = s_M a \sin\theta \frac{\partial \underline{u}^F}{\partial t}\end{aligned} \quad [3.148]$$

The dipole $D_S H(t)$, modeling the step variation in the thermoelastic source, generates the same acoustic field as an impulse force of linear density $s_M D_S \sin\theta\delta(t)$. Radiation by a line source of tangential force density F_1 has been discussed in section 1.2.2.1. According to expression [1.152], the displacement created by the thermoelastic source is:

$$u_M(r,\theta,t) = \frac{D_S}{\pi\mu V_M} A_M(\theta) \frac{H(t - s_M r)}{\sqrt{t^2 - s_M^2 r^2}} \quad \text{with} \quad A_M(\theta) = D_{1M}(\theta)\sin\theta \quad [3.149]$$

Given relations [3.132] and [3.147], the dipole density of a surface thermoelastic source of length l is equal to:

$$D_S = \frac{1-2\nu}{1-\nu} K \frac{\Delta V}{l} = \frac{1-2\nu}{1-\nu} K \frac{3\alpha}{\rho C}\frac{Q}{l} \quad [3.150]$$

ORDER OF MAGNITUDE.– With duralumin constants and an absorbed energy $Q = 1$ mJ over a length $l = 1$ cm, that is, $Q/l = 0.1$ J/m, the dipole density (homogeneous to a force) is $D_S \approx 0.1$ N.

The directivity function of longitudinal waves $A_L(\theta) = D_{1L}(\theta)\sin\theta$ is real, whatever the angle θ with the normal to the surface, because the ratio $\kappa^{-2} = (V_L/V_T)^2$ is higher than 2 for any solid (Figure 3.32(a)). The radiation, which is non-existent along the normal, is maximal in a direction depending on the ratio $\kappa = V_T/V_L$, close to 65° in the case of duralumin. The directivity function of transverse waves $A_T(\theta) = D_{1T}(\theta)\sin\theta$ is real for $\theta < \theta_c = \arcsin\kappa$ and complex for $\theta > \theta_c$. Maximum energy is radiated for an angle close to 30° (Figure 3.32(b)). The emission, which is always zero at the epicenter ($\theta = 0$), is also zero for $\theta = 45°$ due to the cancellation of the term $\cos 2\theta$ in formula [1.154]. The polarities of the main and secondary lobes are opposite.

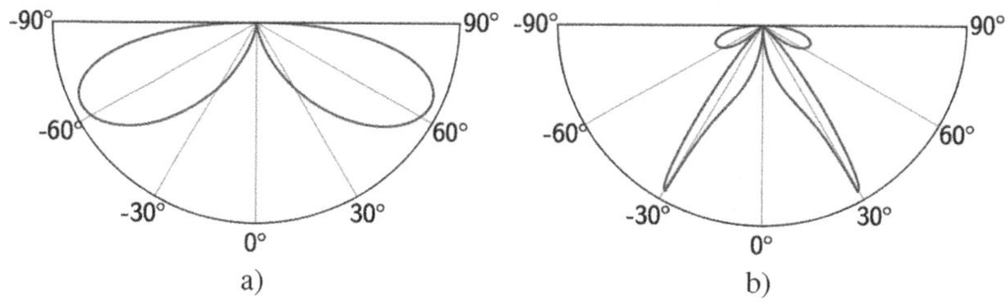

Figure 3.32. *Directivity diagrams of a thermoelastic point or line source for a material with Poisson's ratio $\nu = 1/3$: a) longitudinal wave; b) transverse wave*

A comparison of displacement amplitudes shows that in the thermoelastic regime, transverse waves are generated more efficiently than longitudinal waves. According to formula [3.149], the ratio of the maximum displacements for the two types of waves is equal to:

$$\frac{u_{T_{max}}}{u_{L_{max}}} = \frac{V_L A_{T_{max}}}{V_T A_{L_{max}}} \qquad [3.151]$$

For duralumin ($A_{T_{max}} = 0.84$, $A_{L_{max}} = 0.15$), the transverse displacement emitted by a thermoelastic source is more than one order of magnitude greater than the longitudinal displacement.

The wave fronts approximation is consistent with the far field hypothesis, to the extent that only the high frequency spectral components are considered, so that the observation distance is large compared to the acoustic wavelengths.

3.3.1.3.2. Rayleigh waves

The fact that the thermoelastic source is located closed to the surface, within the material, favors the generation of Rayleigh waves. When the laser pulse is focused along a line of width a, parallel to x_3, Rayleigh waves are preferentially emitted along the x_1 axis perpendicular to the line source. The normal displacement u_2 is the sum of the contributions of forces $+F_1$ and $-F_1$, applied at $\xi_1 = +a/2$ and $-a/2$ (Figure 3.31). On the free surface $x_2 = 0$, the relation [3.148] with $\theta = \pi/2$ shows that the response to a force dipole $D_S H(t)$ is identical to the response to a single impulsive force $s_R D_S \delta(t)$:

$$u_2(x_1, t) = s_R D_S G_{21}^{\delta}(x_1, t) \qquad [3.152]$$

where s_R is the slowness of the Rayleigh wave and $D_S = F_1 a$ the force dipole density [3.150] equivalent to the thermoelastic source on the surface.

According to the results of section 1.2.2.2, the normal displacement of the Rayleigh wave generated by a line source of tangential force density $F_1\delta(t)$ is given by expression [1.177]. In the case of a thermoelastic line source $D_S H(t)$, it is enough to replace F_1 with $s_R D_S$:

$$u_2(x_1, t) = b_{21} s_R D_S \delta(t - s_R x_1) \qquad \text{where} \qquad b_{21} = \frac{R}{4\mu} \qquad [3.153]$$

is the "material" parameter R given by [1.174]. In the case of a dipole density $D_S =$ 0.1 N and a laser pulse with a rectangular profile and time duration $\Delta = 20$ ns, that is, $\delta = 1/\Delta = 5.10^7$ s^{-1}, the instantaneous displacement is of the order of 6.5 nm on a duralumin surface ($V_R = 2\,920$ m/s, $R = 0.384$ and $\mu = 25$ GPa).

In Figure 3.33, the signals measured by the deflection of a laser beam (section 3.4.1.1) on the surface of a titanium alloy (Ti_6Al_4V) are compared to the derivative of the normal displacement $\partial u_2/\partial x_1$. The elastic waves are generated in the thermoelastic regime by a titanium-sapphire laser delivering pulses of duration 100 fs (10^{-13}s). The model, which takes into account thermal diffusion, is similar to the model described in section 1.2 to calculate the displacements emitted in the near field by a line source. The head wave (H), which precedes the Rayleigh wave (R), is clearly visible. The experimental "pump-probe" technique (Higuet *et al.* 2011), which was used to measure the normal displacements, is described in section 3.4.4.

3.3.2. *Ablation regime*

In the field of non-destructive testing, the thermoelastic regime is required to avoid damaging the surface of the sample. Optical detection is difficult due to the small amplitude of the mechanical displacements generated. The ablation regime, which is more efficient, can be used, for example, when the part to be inspected is heated to very high temperatures. On the other hand, the presence of an absorbent film or a transparent layer on the material surface also modifies the efficiency of the thermoelastic source in a favorable way.

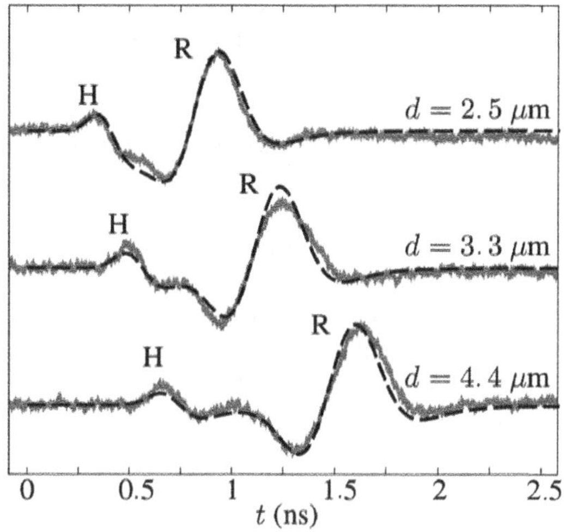

Figure 3.33. *Rayleigh wave (R) generated in the thermoelastic regime on the surface of a titanium alloy (Ti_6Al_4V). The head wave (H) is visible. Comparison between measurements (solid line) and the model (dashed line)*

When the density I of the absorbed light power is sufficient, the laser impact causes melting and then vaporization of a small quantity of matter. The momentum transfer resulting from the ejection of particles creates, in the irradiated zone, a force normal to the surface, which favors the generation of longitudinal waves (Figure 3.27(b)). The ablation regime sets in beyond a threshold (Ready 1971):

$$I > \rho C\sqrt{\frac{\pi D_{th}}{4\Delta}}(T_v - T_i) \qquad [3.154]$$

which depends on vaporization temperature (T_v) and initial temperature (T_i) of the material. For duralumin ($T_v = 2\ 600$ K) with $T_i = 400$ K, and for a laser pulse of duration $\Delta = 20$ ns, ablation occurs if the absorbed power density I is greater than 15 MW/cm^2. Unlike the thermoelastic regime, the absorption coefficient η varies with the incident power density due to the presence of a liquid or gaseous phase on the material surface; it can reach 90%.

The normal force per unit area due to vaporization is given by Ready (1971):

$$F_2 = \frac{I^2}{\rho L_v[L_v + C(T_v - T_i)]} \qquad [3.155]$$

where L_v is the latent vaporization heat of the material. For an absorbed power density $Q = 80$ MW/cm^2 and an aluminum plate ($L_v = 284 \times 10^6$ J.kg^{-1}), this force is equal to 0.28 N/cm^2. Indeed, most of the material is ejected in a liquid phase rather than in a gaseous phase. This liquid phase significantly increases the normal stress in an ablation regime. When the pulse stops, the ablation phenomenon continues until the temperature of material falls below its vaporization temperature.

The power density I can be increased experimentally by reducing the dimensions of the focal spot (Hutchins 1988). At a level just above the ablation threshold, the thermoelastic contribution still plays an important role. Beyond the ablation threshold, the amplitude of the longitudinal displacement rapidly increases with the light intensity I due to the local melting and the increase in the absorption coefficient (Figure 3.34). Subsequently, the amplitude of the longitudinal displacement passes through a maximum before decreasing. This phenomenon is caused by the screening effect of the plasma, which becomes more and more opaque to the laser radiation.

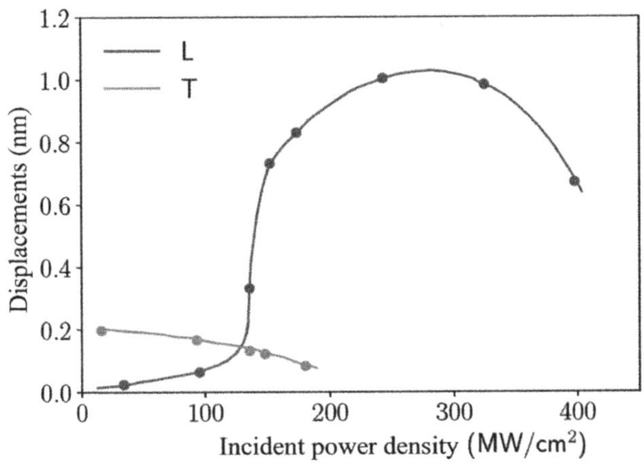

Figure 3.34. *Amplitude of longitudinal (L) and transverse (T) displacements generated in Duralumin versus the incident power density (Dewhurst et al. 1982). For a color version of this figure, see www.iste.co.uk/royer/waves2.zip*

The directivity functions of an ablative source are identical to those of a line of normal forces. In section 1.2.2.1, they have been calculated by the mixed matrix method. For longitudinal waves, $D_{2L}(\theta)$ is given by formula [1.146] and the radiation pattern is shown in Figure 1.20(a). The ablative source is omnidirectional but non-isotropic and the radiation is maximal along the normal to the surface. For transverse waves, $D_{2T}(\theta)$ is given by formula [1.149] and the radiation pattern is represented in Figure 1.20(b). The source is more directional for transverse waves and maximum energy is emitted at an angle close to 35° in the case of duralumin ($\kappa = V_T/V_L = 0.5$).

The amplitude of the *Rayleigh waves* generated in the ablation regime is also increased. Concerning the shape of the signals, in the thermoelastic regime the surface is first displaced outward, while in the ablation regime it is first displaced inward.

3.4. Optical measurement of mechanical displacements

This section aims to describe optical instruments suitable to measure mechanical displacements perpendicular to a surface and produced by elastic waves. In the ultrasonic frequency range (100 kHz to 1 GHz), these displacements are generally small, of the order of a fraction of a nanometer. Optical probes are less sensitive than piezoelectric detectors; however, they offer the advantage, *a priori*, of a local examination without mechanical contact and in a very large bandwidth.

The interaction of a light beam, of diameter d, with an acoustic wave of wavelength λ, modifies:

– the wave direction if $d < \lambda$; the light beam is deflected by the undulation of the surface;

– the wave intensity if $d >> \lambda$; the light beam that covers several wavelengths is diffracted by the phase grating associated with the elastic wave;

– the wave phase through the variation in the optical path caused by the normal displacement of the surface;

– the wave frequency through the Doppler effect.

Optical detectors of mechanical displacements are grouped here into two categories: the non-interferometric probes that exploit the deflection or diffraction of the light beam and the interferometric probes that exploit the phase or frequency modulation of the light wave.

3.4.1. *Non-interferometric methods*

These are used for studying a wave that gives rise to a surface corrugation, such as a Rayleigh wave. This wave propagates on the surface of any solid over a depth of the order of the wavelength (Volume 1, Chapter 3).

3.4.1.1. *Deflection*

Figure 3.35 illustrates the operating principle of this technique. The light beam, reflected by the surface, oscillates as the Rayleigh wave passes. It is partially masked by a wedge (in practice, the edge of the photodetector) so that the intensity of the photocurrent is modulated at the wave frequency.

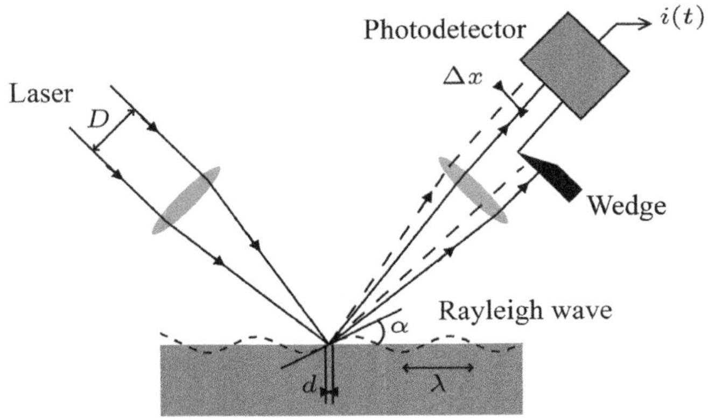

Figure 3.35. *Deflection of a light beam by a Rayleigh wave. On the surface, the dimension d of the light beam is small compared to the wavelength λ_R. The direction of the reflected beam oscillates at the wave frequency*

Let $u = u_0 \sin(\omega t - kx)$ be the mechanical displacement normal to the surface, the maximum slope of the surface undulation is $\alpha = ku_0$ and the amplitude of the angular deviation of the reflected beam is 2α. After the second lens with a focal length F (assumed to be equal to that of the first lens), the displacement of the beam is $\Delta x = 2\alpha F$. If half the beam, of diameter D and power P_0, is masked by the wedge, the intensity of the current $i(t)$ delivered by the photodetector, of response factor S (in A/W), varies as:

$$i(t) = \frac{\Delta x}{D} S P_0 = 2ku_0 \frac{F}{D} S P_0 \sin(\omega t - kx) \qquad [3.156]$$

Assuming that the diameter d of the focal spot on the surface is limited by the diffraction: $d = \Lambda F/D$, where Λ is the optical wavelength and denoting by $I_0 = SP_0/2$ the continuous part of the current intensity, we get:

$$i(t) = 4ku_0 \frac{d}{\Lambda} I_0 \sin(\omega t - kx) \qquad [3.157]$$

These considerations are valid as long as $d < \lambda$. If the focal spot covers several elastic wavelengths, the beam is diffracted (see the following section). For the optimal value $d = \lambda/2$ and since $k\lambda = K\Lambda = 2\pi$, we get:

$$i_{opt}(t) = 2Ku_0 I_0 \sin(\omega t + \varphi) \qquad [3.158]$$

where K is the optical wave number.

This technique lends itself well to wave front scanning. It has the disadvantage of requiring a very good surface state and to be limited in frequency: the maximum value, defined by $d = \lambda$, is equal to $DV/\Lambda F$. For $F = 10$ cm, a laser beam of diameter $D = 1$ mm and a surface wave velocity $V = 3\ 000$ m/s, the maximum frequency is equal to 47 MHz. However, frequencies in the GHz range can be detected by using a beam expander to increase the diameter D and a microscope objective to reduce the focal length F (see Figures 3.33 and 3.42). This easy-to-implement method is used in other fields, for example, to detect the oscillation of a very small cantilever in an atomic force microscope.

3.4.1.2. *Diffraction*

When the light beam covers several wave fronts ($d >> \lambda$), the surface corrugation acts as a phased array. For an angle of incidence θ_0, the beam is diffracted into beams with frequencies $\Omega \pm m\omega$, along directions such that (Royer and Dieulesaint 1999):

$$\sin\theta_m = \sin\theta_0 \pm m\Lambda/\lambda \qquad \text{with} \qquad m = 0, 1, \text{etc.} \tag{3.159}$$

If the surface is perfectly reflective, the power P_m of the diffracted beam of order m is given by the square of the Bessel function $J_m(\Delta\Phi)$, whose argument is equal to the phase shift $\Delta\Phi = 2Ku_0\cos\theta_0$ produced by the displacement normal to the surface, of amplitude u_0:

$$\frac{P_m}{P_0} = \left[J_m(2Ku_0\cos\theta_0)\right]^2 \tag{3.160}$$

When $Ku << 1$, $J_1(\Delta\Phi) \approx \Delta\Phi/2$: the intensity of the first-order diffracted beam is proportional to the square of the mechanical displacement:

$$\frac{P_1}{P_0} \approx (Ku_0\cos\theta_0)^2 \tag{3.161}$$

This displacement can be deduced from the measurement of the relative intensity of this beam. Since diffracted beams must be separated, the method is only applicable at high frequencies ($f > 100$ MHz). The sensitivity to small displacements is low because the photocurrent intensity is proportional to $(Ku_0)^2$. For example, with a He-Ne laser ($\Lambda = 633$ nm), $Ku_0 \approx 10^{-2}$ for $u_0 = 1$ nm. This simple technique is useful for mapping in a steady-state regime the spatial distribution of power for a beam of elastic surface waves. It does not provide the phase of the elastic wave, which is contained in the phase of the diffracted light beam. To find this, the diffracted beam must be mixed with a part of the incident beam.

3.4.2. *Interferometric methods*

These probes are suitable for measuring any mechanical displacement normal to a surface, regardless of its origin (Rayleigh wave, Lamb wave, bulk wave, or other vibration or shock) in a steady-state regime as well as in a transient regime. The phase Φ of a light beam reflected by the vibrating surface of an object is modulated by the normal displacement $u_0 \sin(\omega t + \varphi)$ of this surface:

$$\Delta\Phi = 2Ku_0 \sin(\omega t + \varphi) \qquad [3.162]$$

Since optical detectors are only sensitive to light intensity, to be exploitable this phase modulation must be converted either into an amplitude modulation using a homodyne interferometer (of Michelson type, for example) or transposed into the radio-frequency domain using a heterodyne interferometer. When the frequency f is high enough, greater than a few megahertz, the information on the mechanical displacement can be, *a priori*, directly extracted through optical spectroscopy with the help of a Fabry-Pérot cavity. Each of these three methods will be studied in turn.

3.4.2.1. *Homodyne probe, stabilized Michelson interferometer*

The first method consists of mixing the probe beam S reflected by the object, with a reference beam R emanating from the same source, of power P_0. Let E_S and E_R be the electric fields of the two beams, assumed to have the same amplitude $E_0/\sqrt{2}$, corresponding to half-power $P_0/2$:

$$\begin{cases} E_S = \dfrac{E_0}{\sqrt{2}} \exp i(\Omega t + \Delta\Phi + \Phi_S) \\[2ex] E_R = \dfrac{E_0}{\sqrt{2}} \exp i(\Omega t + \Phi_R) \end{cases} \qquad [3.163]$$

The photodetector provides a current whose intensity is proportional to the light power received, that is, to the square of the total electric field:

$$I = S(E_R + E_S)(E_R + E_S)^\star = S(|E_R|^2 + |E_S|^2 + 2\,\mathrm{Re}[E_S E_R^\star]) \qquad [3.164]$$

The photocurrent is composed of a continuous part $I_0 = S|E_0|^2$ and an alternative part that contains the displacement of the acoustic wave:

$$I = I_0 + I_0 \cos[2Ku_0 \sin(\omega t + \varphi) + \Phi_S - \Phi_R] \qquad [3.165]$$

This expression highlights the significance in interferometry of fluctuations in the phase difference $\Phi = \Phi_S - \Phi_R$, that is, fluctuations in the difference ΔL between the optical paths L_S and L_R of the two beams. The variations $\Delta\Phi = 2\pi\Delta L/\Lambda$ must be small compared to Ku_0 (it must be recalled that $Ku_0 \approx 10^{-2}$ for $u_0 = 1$ nm).

The most common optical configuration is similar to that of a conventional interferometer (Figure 3.36(a)). The laser beam is divided into two equal parts, which reflect off the object (probe beam S) and off the mirror (reference beam R) before mixing on the photodiode. An insulator prevents the part of the beam that returns toward the source from penetrating into the laser cavity and generating instabilities. For reducing the effect of phase fluctuations, the mirror reflecting the reference beam is displaced in order to maintain constant the path difference $L_S - L_R$ between the light beams (stabilized Michelson interferometer).

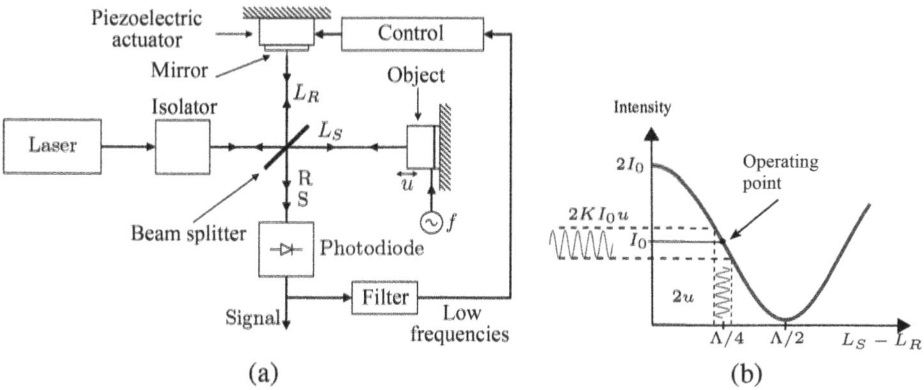

Figure 3.36. *Stabilized Michelson interferometer. a) The position of the reference mirror is controlled by the "low frequency" part of the photodiode output signal, in order to maintain the operating point fixed, despite fluctuations in the optical paths. b) Variation in the photocurrent intensity as a function of the optical path difference. The sensitivity is maximum when this difference is equal to* $\Lambda/4$

The photocurrent intensity [3.165] is a sine function of the optical path difference $L_S - L_R$, with period Λ (Figure 3.36(b)). The sensitivity is maximum for:

$$L_S - L_R = (n \pm 1/4)\Lambda \qquad \text{hence} \qquad \Phi_S - \Phi_R = \pm\pi/2 \qquad [3.166]$$

The position of the reference mirror is controlled by a feedback loop, so that the operating point is maintained in this phase quadrature position, despite random fluctuations in optical paths (Kwaaitaal 1974). The mirror is carried by a piezoelectric actuator controlled by the low-frequency part ($f < 1$ kHz) of the detected signal, which contains most of the spectrum of thermal and mechanical disturbances. However, the limited dynamic range of the feedback loop requires periodic adjustments.

At this operating point in phase quadrature, and when the displacements are small ($Ku_0 << 1$), the photocurrent intensity is given by:

$$I = I_0 + 2Ku_0 I_0 \sin(\omega t + \varphi) = I_0 + i_S(t) \qquad [3.167]$$

The sensitivity is limited by the shot noise (or photon noise) arising from the continuous component I_0. This generates a current i_N whose mean quadratic intensity is proportional to the bandwidth B of the electronic detection circuit:

$$\langle i_N^2(t) \rangle = 2eI_0 B, \qquad e: \text{electron charge} \qquad [3.168]$$

The smallest measurable displacement u_{min}, defined by a signal-to-noise ratio equal to unity, is written as:

$$\frac{S}{N} = \sqrt{\frac{\langle i_S^2(t) \rangle}{\langle i_N^2(t) \rangle}} = Ku_0 \sqrt{\frac{I_0}{eB}} \quad \text{hence} \quad u_{min} = \frac{\Lambda}{2\pi} \sqrt{\frac{eB}{I_0}} \qquad [3.169]$$

For a He-Ne laser, a power $P_0 = 1$ mW received by the photodiode ($I_0 = 0.3$ mA) and a bandwidth $B = 1$ Hz, we get:

$$u_{min} = 2.3 \times 10^{-6} \text{ nm}/\sqrt{\text{Hz}} \qquad [3.170]$$

that is, 10^{-2} nm for a bandwidth of 20 MHz. This theoretical limit is only reached if the thermal noise in the load resistance R of the photodiode: $4k_B TB/R$ (k_B is the Boltzmann constant) is less than the shot noise. In practice, this condition $R >> 2k_B T/eI_0$, or $R >> 150\ \Omega$, is all the more difficult to fulfill the higher the frequency. This is due to parasitic capacitances that reduce the bandwidth.

3.4.2.2. *Heterodyne probe*

In a heterodyne interferometer, the frequency of one (or both) beam(s) is shifted. This change in frequency, equal to $\pm f_B$, distinguishes this type of interferometer from the previous one of homodyne type. It is implemented by an acousto-optic modulator (Royer and Dieulesaint 1999), in which the frequency of the elastic waves (f_B) is of several tens of MHz. The electric fields E_S and E_R are given by expressions [3.163] upon replacing Ω by Ω_S for the probe beam, and by Ω_R for the reference beam. A term at angular frequency $\Omega_S - \Omega_R = \omega_B$ appears in the alternative part of the current, supplied by the photodetector (equation [3.165]):

$$i_S(t) = I_0 \cos\left[\omega_B t + 2Ku_0 \sin(\omega t + \varphi) + \Phi_S - \Phi_R\right] \qquad [3.171]$$

The phase modulation of the probe beam by the surface displacement is thus transposed into the radio-frequency domain. The spectrum of $i_S(t)$ includes a central line at ω_B and lateral lines at $\omega_B \pm m\omega$, whose heights are given by Bessel functions $J_m(2Ku_0)$. This result can be established by using the complex representation and Jacobi–Anger identity ([A1.18], Appendix A1.1), with $x = 2Ku_0$ and $\theta = \omega t + \varphi$. If the displacement u_0 is small compared to the optical wavelength ($Ku_0 << 1$), that is, practically, $u_0 < 20$ nm:

$$J_0(2Ku_0) \approx 1, \quad J_1(2Ku_0) \approx Ku_0, \quad J_m(2Ku_0) \approx 0 \text{ for } m \geq 2 \qquad [3.172]$$

the spectrum is reduced to the carrier frequency ω_B and to the lateral lines $\omega_B \pm \omega$ (Figure 3.37(b)):

$$\begin{aligned} i_S(t) &= I_0 \{\cos(\omega_B t + \Phi_S - \Phi_R) + Ku_0 \cos[(\omega_B + \omega)t + \varphi + \Phi_S - \Phi_R] \\ &\quad - Ku_0 \cos[(\omega_B - \omega)t - \varphi + \Phi_S - \Phi_R]\} \end{aligned} \qquad [3.173]$$

The ratio $r = 1/Ku_0$ of the heights of the central line and lateral lines provides the absolute amplitude of the mechanical vibration in the steady-state regime, independently of the light power reflected by the surface. For a He-Ne laser and without any calibration procedure, the normal displacement amplitude is given by:

$$u_0 = \frac{\Lambda_0}{2\pi}\frac{1}{r} \approx \frac{100}{r} \text{ (nm)} \qquad \text{with} \qquad \Lambda_0 \approx 633 \text{ nm} \qquad [3.174]$$

3.4.2.2.1. Compact optical configuration

A Mach–Zehnder type structure, like the arrangement shown in Figure 3.37(a), may be made very compact (Royer and Dieulesaint 1986). The source beam, horizontally polarized, is divided by the beam splitter cube A into a reference beam (R) and a probe beam (S). The first one is directed by the Dove prism P toward the photodiode. The second one, whose frequency is shifted by f_B (70 MHz) by the acousto-optic modulator (AOM), with collinear input and output, is reflected by the sample, which vibrates at frequency f. Having passed across the quarter-wavelength plate twice, it comes back vertically polarized to be reflected, at 90°, by the polarization beam splitter B toward the photodiode. The two beams, R and S, whose frequencies are F and $F + f_B$, respectively, pass through an analyzer oriented at 45° with respect to their respective polarizations, before interfering on the photodiode. This one delivers a current at frequency f_B, whose phase is modulated by the vibration of the sample surface. Figure 3.37(b) depicts the photocurrent spectrum for a mechanical displacement equal to 1 nm at 100 kHz.

The lens focusing the probe beam onto the object improves the spatial resolution, increases the amount of light collected in the case of a diffusing surface and makes the probe less sensitive to the tilt of the sample when it is displaced. The shift from the axis of the two cubes eliminates the spurious signals arising from the reflections of the S and R beams (at points a and b) at the interface of the polarization beam splitter B.

This probe can be rapidly aligned because the modulator is only crossed once and only by the probe beam. Moreover, the two beams are either parallel or perpendicular.

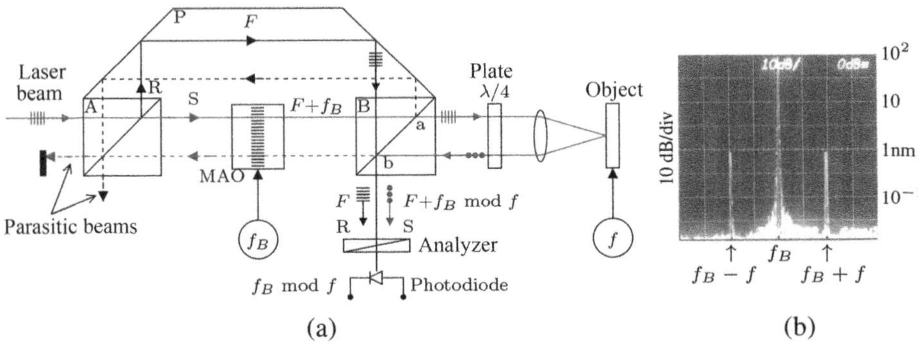

Figure 3.37. *Heterodyne interferometric probe. a) Compact optical arrangement. b) Spectrum of the photodiode output signal for a surface displacement of one nanometer at 100 kHz. The noise level shows that a displacement lower than 0.03 nm can be read directly on a spectrum analyzer (bandwidth 3 kHz)*

3.4.2.2.2. Coherent electronic detection

Formula [3.173] indicates that random phase fluctuations $\Phi_S - \Phi_R$ affect the central line and lateral lines of the spectrum in the same way. In principle, they can be eliminated by a coherent electronic detection, which makes the difference between the frequencies of the central line and one of the lateral lines. Such detection can be implemented according to various ways. Figure 3.38 represents a wideband processing suitable for detecting transient displacements. The signal that remains after elimination of the "acoustic" phase modulation by a narrow-passband filter, centered on the angular frequency ω_B, is phase shifted by $\pi/2$:

$$i_F(t) \propto \cos\left(\omega_B t + \pi/2 + \Phi_S - \Phi_R\right) \qquad [3.175]$$

The component at frequency $2\omega_B$ is eliminated by a low-pass filter, then the filtered signal is mixed with the photocurrent $i_S(t)$, given by the formula [3.171], so that:

$$s(t) = i_S(t).i_F(t) \propto I_0 \cos\left[\frac{\pi}{2} - 2Ku_0 \sin(\omega t + \varphi)\right] \qquad [3.176]$$

The output signal is proportional to the instantaneous mechanical displacement when $Ku_0 << 1$:

$$s(t) \approx 2Ku_0 I_0 \sin(\omega t + \varphi) \qquad [3.177]$$

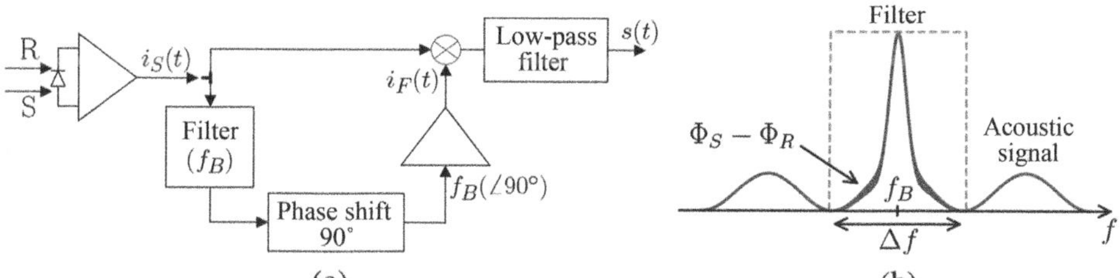

Figure 3.38. *Broadband coherent detection of the heterodyne probe. a) The mixing of the part of the photocurrent at the carrier frequency f_B, phase shifted by $\pi/2$, with the other, unmodified part, yields, after filtering, the vibration of the object. b) In order to eliminate the fluctuations by difference, the bandwidth of the filter, centered at the carrier frequency f_B, is larger than their spectrum*

It should be noted that a phase loop circuit is appropriate to measure surface displacements of very low frequency ($f < 1$ kHz).

3.4.3. *Doppler velocimetry*

In the interferometers (whether homodyne or heterodyne) described above, the reflected wave, whose wave surface may be strongly disturbed, is mixed with a quasi-plane reference wave. In the case of diffusing surfaces, that is, in the presence of *speckle*, these probes can only collect a grain of light, coherent in phase: their geometrical *étendue* is limited. To overcome this problem, only a single wave must be used, namely the wave originating from the surface whose displacement is to be measured. The information about the movement results from the interference (or beating) between a part of this wave and another part delayed by the time τ. The distortion of the wave fronts during reflection is not significant because, in a first approximation, it affects both waves in the same way. As shown in Figure 3.39, the interference can be achieved by a Michelson interferometer and the delay by a multiple reflection device such as a Fabry-Pérot cavity (Monchalin 1986) or by an optical fiber.

By setting $\tau = L/c$, the phase shift due to the difference L between optical paths "1" and "2" is $\Delta\Phi = \Omega\tau$. If $2Ku(t)$ is, for beam "1", the phase shift produced by the surface displacement at time t, the phase shift undergone, at the same time, by beam "2" is $2Ku(t-\tau)$. According to formula [3.165], the photocurrent resulting from the beating of the two beams is:

$$I = I_0 + I_0 \cos\{2K[u(t) - u(t-\tau)] + \Omega\tau + \phi\} \qquad [3.178]$$

where ϕ is an adjustable phase shift introduced into one of the two paths, in order to maintain the interferometer in phase quadrature condition:

$$\Omega\tau + \phi = \pm\pi/2 \qquad (modulo \quad 2\pi) \tag{3.179}$$

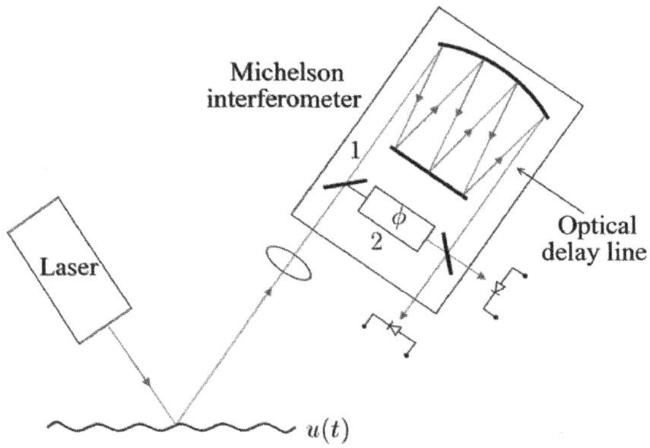

Figure 3.39. *Principle of time-delay differential interferometry*

For a displacement of amplitude $u \ll \Lambda$, the variable component of the photocurrent is given by:

$$i(t) = I_0 \sin\{2K[u(t) - u(t-\tau)\} \approx 2KI_0[u(t) - u(t-\tau)] \tag{3.180}$$

This result can be exploited in different ways:

– if the delay τ is large compared to the time duration Θ of the signal: $u(t-\tau) = 0$ when $u(t) \neq 0$, the current intensity is proportional to the normal displacement of the surface;

– if, on the contrary, the delay τ is small compared to Θ: $u(t) - u(t-\tau) \approx \tau\, du/dt$, the signal is proportional to the normal component of the particle velocity $v(t) = du/dt$. The device then operates as a Doppler velocimeter. After reflection by the moving surface, the incident wave $E_0 \exp(i\Omega t)$ becomes $E_1 \exp i[\Omega t + 2Ku(t)]$. Its instantaneous angular frequency, that is, the time derivative of the phase:

$$\Phi_1 = \Omega t + 2Ku(t) \qquad \longrightarrow \qquad \Omega_1 = \frac{d\Phi_1}{dt} = \Omega + 2\Omega\frac{v(t)}{c} \tag{3.181}$$

is shifted with respect to the angular frequency Ω of the incident light wave. This Doppler effect appears in the spectrum of the reflected wave in the form of lateral bands, as seen in formula [3.173], which is valid for a sinusoidal displacement $u(t)$. In the instrument developed by Monchalin, this frequency variation is transformed into an amplitude variation by a confocal Fabry-Pérot cavity of length 50 cm (Monchalin 1986). The cavity bandwidth is 1.5–10 MHz depending on the reflectivity of the mirrors. Its length is controlled by piezoelectric actuators (Figure 3.40(a)), in order to maintain the laser frequency at the mid-height of the transmission curve (operating as a frequency discriminator as shown in Figure 3.40(b)).

This instrument is able to detect mechanical displacements generated by laser pulses in solids heated to high temperatures or presenting a surface of poor optical quality (for example, composite materials). An industrial ultrasonic system (LUIS) has been developed for the inspection of structural elements in aeronautics.

Optical methods described above are particularly well-suited to the experimental study of phenomena giving rise to resonances. Indeed, the absence of mechanical contact prevents any perturbation and damping of these resonances, as it would be the case with piezoelectric transducers.

The example chosen to illustrate these possibilities is the local resonance phenomenon due to Lamb modes with zero group velocity (section 4.3.1.4, Volume 1). The experimental setup comprises a YAG laser delivering pulses of 20 ns duration, at a rate of 50 Hz. Given the size of the spot (the order of 1 mm) and the energy of the laser pulse (a few mJ), the Lamb waves are generated in the thermoelastic regime, that is, without any alteration of the surface of the sample. The mechanical displacement is measured at the point of the laser impact on a duralumin plate (of thickness $d = 0.49$ mm) by a heterodyne optical probe (see section 3.4.2.2).

Figure 3.41(a) shows the signal recorded during the 100 μs following the impact of the laser pulse. The low-frequency oscillations observed in the first 10 μs correspond to the excitation of the flexural mode A_0. Those between 45 and 85 μs are due to the reflections of this mode from the edges of the plate (of lateral dimensions 100 $\times$ 150 mm^2). In the interval, a high-frequency component appears, whose spectral content is analyzed in Figure 3.41(b). The spectrum is dominated by a resonance at a frequency (5.86 MHz) very close to the theoretical value (5.85 MHz) for the minimum of the dispersion curve of the first symmetric Lamb mode S_1 (Figure 4.15, Volume 1). The small peak at 9.6 MHz corresponds to the A_2/A_3 ZGV mode, whose frequency $\times$ thickness product is $fd = 4.71$ MHz.mm.

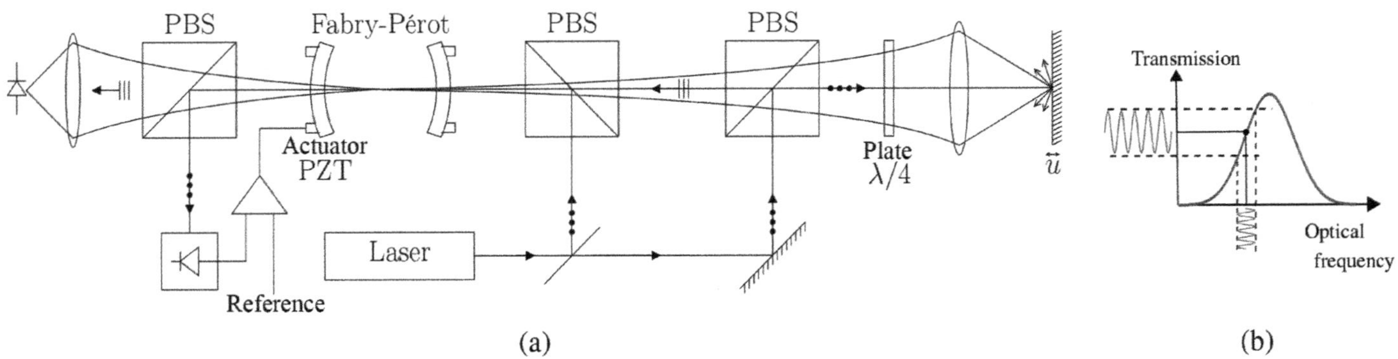

Figure 3.40. *a) Fabry-Pérot interferometer with a confocal cavity (Monchalin 1986); PBS denotes a polarizing beam splitter. b) The position of one mirror of the cavity is controlled so that the optical frequency is maintained on the slope of the transmission curve*

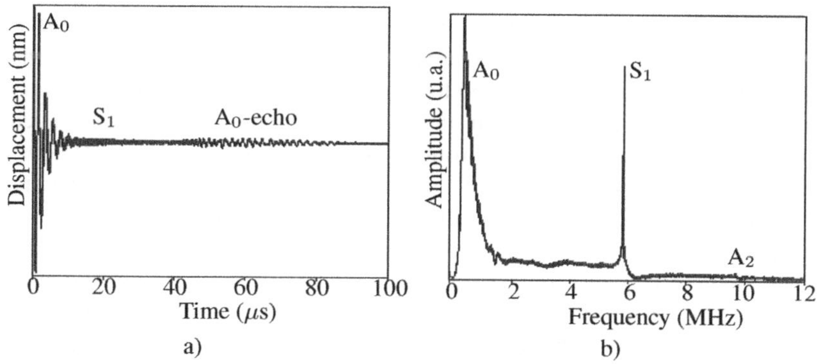

Figure 3.41. *a) Normal displacement generated by a YAG laser pulse on a duralumin plate of thickness 0.49 mm and detected at the same point by a heterodyne optical interferometer. b) Spectrum of the signal*

3.4.4. *Picosecond acoustics*

In the laser-ultrasound methods described above, optical generation and detection are independent: two lasers and two different instruments are used. The pulse duration lies between a few ns and a few tens of ns, the frequency of the elastic waves varies from a few MHz to a few hundred MHz and the waves travel a distance ranging from a few mm to tens of cm. The emergence, in the 1980s, of titanium-sapphire type lasers have modified these orders of magnitude and opened up new possibilities for the investigation of matter on a much smaller scale. These lasers emit pulses of a duration between 50 fs and a few picoseconds, at a rate of about 80 MHz. The frequency of the elastic waves generated by the thermoelastic effect (section 3.3.1.1) goes beyond the GHz and, taking into account the attenuation, the distances covered are micrometric. The propagation times of elastic waves, of the order of the nanosecond, can easily be compensated by an optical delay line (at the speed of light, 1 ns is equivalent to 30 cm in the air).

In the diagram in Figure 3.42, which is typical of a picosecond acoustics experiment, the laser beam is divided into two parts by the polarizing beam splitter A. The pump beam is focused onto the surface of the sample, while the probe beam passes through a motorized variable delay line before also being focused onto the surface. The reflected beam is directed toward the photodetector by the combination of the quarter-wavelength plate and the polarizing beam splitter B.

In its original version, developed by H. Maris (Thomsen *et al.* 1984) and his team, the probe and pump beams are focused at the same point and the quantity that is detected is the variation in reflectivity, produced by the longitudinal wave train, reflected by layers located below the surface of the sample. The absorption of a part of the light energy causes a sudden variation in the reflectivity of the surface. The thermalization of electrons leads to a regular decrease in the thermal signal. The dilatation accompanying the reflected longitudinal elastic waves induces a very small relative variation (typically 10^{-4}) in the surface reflectivity. To measure such small signals, the pump beam passes through an acousto-optic device (AOM), which modulates the intensity of the femto- or picosecond pulse trains at a frequency of a few hundred kHz. The current delivered by the photodiode is filtered at this frequency, with a bandwidth of the order of 1 Hz, and is then amplified by the lock-in amplifier. If the reflected acoustic wave coincides with the light pulse of the probe beam, a signal is detected. The coincidence is obtained by shifting the mirrors of the delay line by a distance such that the optical delay is equal to the acoustic travel time. It is, therefore, possible to analyze the sample at different depths. This technique was developed in recent years to study the mechanical properties of thin films and multi-layer structures at the nanometric scale. Such devices are used in the micro-electronic industry for the non-destructive evaluation of materials.

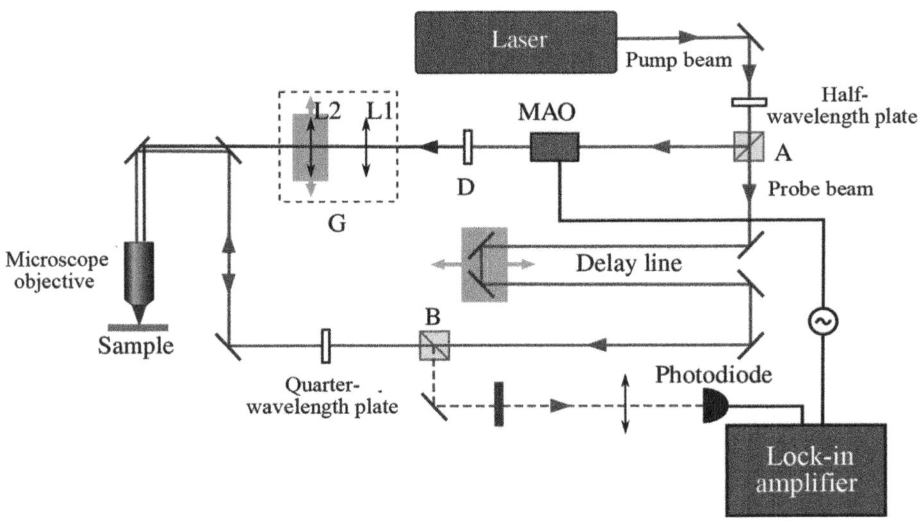

Figure 3.42. *"Pump-probe" setup used in picosecond acoustics. The energy partition between pump and probe beams is adjusted by the half-wavelength plate and the polarization splitter A. After the second passage in the quarter-wavelength plate, the polarization of the probe beam has rotated from 90° and is then reflected by the beam splitter B and directed toward the photodetector. A beam expander (G) and an optical frequency doubler (D) improve the spatial resolution*

One variant consists of separating the source and probe points by a few μm on the sample surface and detecting the Rayleigh waves emitted by the laser pulse. Several detection methods can be implemented: interferometry (section 3.4.2), deflection of the reflected beam (section 3.4.1.1) or the periodic variation of its diameter due to the local curvature of the surface (Higuet *et al.* 2011). For example, the signals in Figure 3.33 were obtained by deflectometry.

Appendix 1

Developments in Cylindrical and Spherical Harmonics

The use of cylindrical or spherical coordinates simplifies the resolution of equations that govern propagation in structures with axial or spherical symmetry. In this appendix, we establish the solutions to the Helmholtz equations in these two curvilinear coordinate systems.

A1.1. Cylindrical harmonics

Let R_c be the reference frame of cylindrical coordinates (r, θ, z) and center O. The Cartesian coordinates are then defined in terms of the cylindrical coordinates by the relations:

$$x = r\cos\theta \quad \text{and} \quad y = r\sin\theta \qquad \text{[A1.1]}$$

Considering expression [A1.7] for the Laplacian in cylindrical coordinates given in Appendix 1 of Volume 1, the Helmholtz equation is written with M = L or T, depending on the type of wave:

$$\left[\frac{\partial^2}{\partial r^2} + \frac{1}{r}\frac{\partial}{\partial r} + \frac{1}{r^2}\frac{\partial^2}{\partial \theta^2} + \frac{\partial^2}{\partial z^2} + k_M^2\right]\phi(r, \theta, z) = 0 \qquad \text{[A1.2]}$$

The solution to this equation is found using the separation of variables method:

$$\phi(r, \theta, z) = R(r)\Theta(\theta)Z(z) \qquad \text{[A1.3]}$$

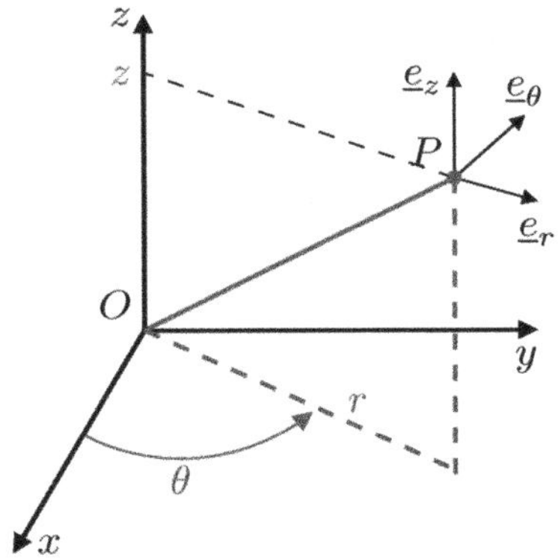

Figure A1.1. *Cylindrical coordinate system*

Substituting this trial solution into equation [A1.2] and dividing by $R\Theta Z$ leads to:

$$\frac{1}{R}\left(\frac{\mathrm{d}^2R}{\mathrm{d}r^2}+\frac{1}{r}\frac{\mathrm{d}R}{\mathrm{d}r}\right)+\frac{1}{r^2}\frac{1}{\Theta}\frac{\mathrm{d}^2\Theta}{\mathrm{d}\theta^2}+\frac{1}{Z}\frac{\mathrm{d}^2Z}{\mathrm{d}z^2}+k_M^2=0 \qquad [A1.4]$$

The third term is a function of the variable z only, while the rest depends on r and θ. It is necessarily equal to a constant $(-k^2)$, leading to equation:

$$\frac{\mathrm{d}^2Z}{\mathrm{d}z^2}+k^2Z=0 \qquad [A1.5]$$

whose solutions are of the type:

$$Z(z)=a_z e^{ikz}+b_z e^{-ikz} \qquad [A1.6]$$

The same reasoning applied to the second term of equation [A1.4], which is written as:

$$\frac{r^2}{R}\left(\frac{\mathrm{d}^2R}{\mathrm{d}r^2}+\frac{1}{r}\frac{\mathrm{d}R}{\mathrm{d}r}\right)+\frac{1}{\Theta}\frac{\mathrm{d}^2\Theta}{\mathrm{d}\theta^2}+\left(k_M^2-k^2\right)r^2=0 \qquad [A1.7]$$

leads to the equation:

$$\frac{1}{\Theta}\frac{\mathrm{d}^2\Theta}{\mathrm{d}\theta^2}=-n^2 \qquad \text{hence} \qquad \frac{\mathrm{d}^2\Theta}{\mathrm{d}\theta^2}+n^2\Theta=0 \qquad [A1.8]$$

where n is a constant. The solutions of this equation are:

$$\Theta_n(\theta) = a_n^\theta e^{in\theta} + b_n^\theta e^{-in\theta} \quad \text{[A1.9]}$$

Since the domain of study is defined for $\theta \in [0, 2\pi]$, n is necessarily an integer, so that the field is continuous at $\theta = 0$ and at $\theta = 2\pi$.

After the substitution, equation [A1.7] takes the form of the Bessel equation:

$$\frac{\mathrm{d}^2 R}{\mathrm{d}r^2} + \frac{1}{r}\frac{\mathrm{d}R}{\mathrm{d}r} + \left(k_\perp^2 - \frac{n^2}{r^2}\right) R = 0 \qquad \text{with} \qquad k_\perp^2 = k_M^2 - k^2 \quad \text{[A1.10]}$$

whose general solution is:

$$R_n(k_\perp r) = a_n^r J_n(k_\perp r) + b_n^r Y_n(k_\perp r) \quad \text{[A1.11]}$$

where J_n and Y_n are Bessel functions of the first and second kind, respectively. The solution $R_n(k_\perp r)$ can also be expressed by:

$$R_n(k_\perp r) = c_n^r H_n^{(1)}(k_\perp r) + d_n^r H_n^{(2)}(k_\perp r) \quad \text{[A1.12]}$$

where $H_n^{(1)}$ and $H_n^{(2)}$ are the Hankel functions defined by:

$$H_n^{(1)}(x) = J_n(x) + iY_n(x) \qquad \text{and} \qquad H_n^{(2)}(x) = J_n(x) - iY_n(x) \quad \text{[A1.13]}$$

The derivatives of the Bessel and Hankel functions verify the recurrence relations:

$$\frac{\mathrm{d}F_n(x)}{\mathrm{d}x} = \frac{n}{x} F_n(x) - F_{n+1}(x) = \frac{1}{2}\left[F_{n-1}(x) - F_{n+1}(x)\right] \quad \text{[A1.14]}$$

For $x = 0$, the function J_0 takes the value 1, while the other J_n functions are zero and the Y_n functions are all singular. The Hankel functions are also singular at the origin and have the specific feature that in the far field, they approach the functions:

$$\lim_{x \to \infty} H_n^{(1)}(x) = \sqrt{\frac{2}{\pi x}} e^{i\left(x - \frac{n\pi}{2} - \frac{\pi}{4}\right)}$$

$$\text{and} \quad \lim_{x \to \infty} H_n^{(2)}(x) = \sqrt{\frac{2}{\pi x}} e^{-i\left(x - \frac{n\pi}{2} - \frac{\pi}{4}\right)} \quad \text{[A1.15]}$$

The potential is sought in the form of an expansion on the cylindrical harmonics:

$$\phi(r, \theta, z) = \left[a_z e^{ikz} + b_z e^{-ikz}\right] \sum_{n=-\infty}^{\infty} R_n(k_\perp r) e^{in\theta} \quad \text{[A1.16]}$$

If the problem is independent of the z coordinate, the wave number k is zero and $k_\perp = k_M$; the potential of the mode M is then expressed by:

$$\phi_M(r,\theta) = \sum_{n=0}^{\infty} a_n R_n(k_M r) e^{in\theta} \qquad \text{[A1.17]}$$

The developments of the exponential of a trigonometric function, given by Jacobi and Anger:

$$e^{ix\sin\theta} = \sum_{n=-\infty}^{n=+\infty} J_n(x) e^{in\theta} \quad \text{and} \quad e^{ix\cos\theta} = \sum_{n=-\infty}^{n=+\infty} i^n J_n(x) e^{in\theta} \qquad \text{[A1.18]}$$

can be used, for example, to represent a phase-modulated wave or to convert a plane wave into a sum of cylindrical harmonics. These identities are deduced from the generating functions of the Bessel functions:

$$e^{\frac{x}{2}\left(t-\frac{1}{t}\right)} = \sum_{n=-\infty}^{n=+\infty} J_n(x) t^n \quad \text{with} \quad t = e^{i\theta} \quad \text{or} \quad t = ie^{i\theta} \qquad \text{[A1.19]}$$

Since $i^{-n} J_{-n}(x) = i^n J_n(x)$, the second equation [A1.18] takes the form:

$$e^{ix\cos\theta} = \sum_{n=0}^{\infty} \epsilon_n i^n J_n(x)\cos(n\theta) \quad \text{by writing} \quad \epsilon_n = 2 - \delta_{n0} \qquad \text{[A1.20]}$$

A1.2. Spherical harmonics

Let R_s be the reference frame of spherical coordinates (r, θ, φ) and center O. The Cartesian coordinates are then defined as a function of the spherical coordinates through the relations:

$$\begin{cases} x = r\sin\theta\cos\varphi \\ y = r\sin\theta\sin\varphi \\ z = r\cos\theta \end{cases} \qquad \text{[A1.21]}$$

Considering expression [A1.16] for the Laplacian in spherical coordinates, given in Appendix 1 of Volume 1, the Helmholtz equation is written in spherical coordinates as:

$$\left[\frac{\partial^2}{\partial r^2} + \frac{2}{r}\frac{\partial}{\partial r} + \frac{1}{r^2\sin\theta}\frac{\partial}{\partial\theta}\left(\sin\theta\frac{\partial}{\partial\theta}\right) + \frac{1}{r^2\sin^2\theta}\frac{\partial^2}{\partial\varphi^2} + k^2\right]\phi(r,\theta,\varphi) = 0$$

[A1.22]

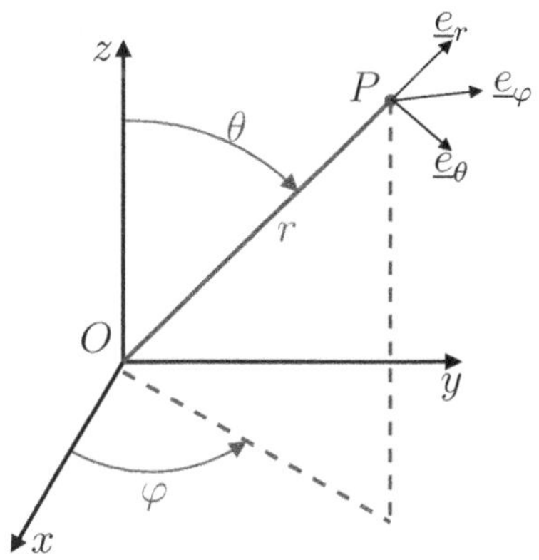

Figure A1.2. *Spherical coordinate system*

The solutions are found using the separation of variables method:

$$\phi(r,\theta,\varphi) = R(r)\Theta(\theta)\Phi(\varphi) \qquad [A1.23]$$

which, after dividing by $R\Theta\Phi$, leads to the equation:

$$\frac{1}{r^2 R}\frac{\mathrm{d}}{\mathrm{d}r}\left(r^2\frac{\mathrm{d}R}{\mathrm{d}r}\right) + \frac{1}{r^2\Theta\sin\theta}\frac{\mathrm{d}}{\mathrm{d}\theta}\left(\sin\theta\frac{\mathrm{d}\Theta}{\mathrm{d}\theta}\right) + \frac{1}{r^2\Phi\sin^2\theta}\frac{\mathrm{d}^2\Phi}{\mathrm{d}\varphi^2} + k^2 = 0 \qquad [A1.24]$$

Multiplying by $r^2\sin^2\theta$ shows that the third term depends only on φ, while the rest depends on r and θ. Consequently, Φ satisfies the following equation:

$$\frac{1}{\Phi}\frac{\mathrm{d}^2\Phi}{\mathrm{d}\varphi^2} = \text{Const.} = -m^2 \qquad \text{hence} \qquad \frac{\mathrm{d}^2\Phi}{\mathrm{d}\varphi^2} + m^2\Phi = 0 \qquad [A1.25]$$

whose solutions are:

$$\Phi_m(\varphi) = a_m^\varphi e^{im\varphi} + b_m^\varphi e^{-im\varphi}, \qquad \text{with} \qquad m \in \mathbb{N} \qquad [A1.26]$$

to ensure continuity over the domain $\varphi \in [0, 2\pi]$.

Considering this solution, equation [A1.24] takes the form:

$$\frac{1}{r^2 R}\frac{\mathrm{d}}{\mathrm{d}r}\left(r^2\frac{\mathrm{d}R}{\mathrm{d}r}\right) + \frac{1}{r^2\Theta\sin\theta}\frac{\mathrm{d}}{\mathrm{d}\theta}\left(\sin\theta\frac{\mathrm{d}\Theta}{\mathrm{d}\theta}\right) - \frac{m^2}{r^2\sin^2\theta} + k^2 = 0 \quad [A1.27]$$

Multiplying by r^2 shows that the first and fourth terms depend only on r, while the second and third terms depend only on θ; equation [A1.27] can be divided into two parts[1]:

$$\begin{cases} \dfrac{1}{\Theta \sin\theta}\dfrac{\mathrm{d}}{\mathrm{d}\theta}\left(\sin\theta\dfrac{\mathrm{d}\Theta}{\mathrm{d}\theta}\right) - \dfrac{m^2}{\sin^2\theta} = C = -n(n+1) \\ \dfrac{1}{R}\dfrac{\mathrm{d}}{\mathrm{d}r}\left(r^2\dfrac{\mathrm{d}R}{\mathrm{d}r}\right) + (kr)^2 = n(n+1) \end{cases} \qquad \text{[A1.28]}$$

By changing the variable $X = \cos\theta$, the θ equation takes the form of an associated Legendre equation:

$$\left(1 - X^2\right)\frac{\mathrm{d}^2\Theta}{\mathrm{d}X^2} - 2X\frac{\mathrm{d}\Theta}{\mathrm{d}X} + \left[n(n+1) - \frac{m^2}{1-X^2}\right]\Theta = 0 \qquad \text{[A1.29]}$$

Its solutions are the *Legendre functions* of degree n and order m:

$$\Theta_{nm}(\theta) = P_n^m(\cos\theta) \qquad \text{with} \qquad -n \leq m \leq n \qquad \text{[A1.30]}$$

which are derived from the Legendre polynomials of degree n (P_n) through a derivative of order m:

$$P_n^m(\cos\theta) = (-\sin\theta)^m\frac{\mathrm{d}^m P_n(\cos\theta)}{\mathrm{d}(\cos\theta)^m} = \frac{\mathrm{d}^m P_n(\cos\theta)}{\mathrm{d}\theta^m} \qquad \text{[A1.31]}$$

If the problem is invariant with respect to the angle φ, $m = 0$ and the solution to equation [A1.29] is a Legendre polynomial of degree n: $P_n(\cos\theta)$.

By performing the change in variable $x = kr$, equation [A1.28] in r is written as:

$$x^2\frac{\mathrm{d}^2R}{\mathrm{d}x^2} + 2x\frac{\mathrm{d}R}{\mathrm{d}x} + \left[x^2 - n(n+1)\right]R = 0 \qquad \text{[A1.32]}$$

The change in function $R = Zx^{-\frac{1}{2}}$ leads to the Bessel equation [A1.10] for the index $n + 1/2$:

$$x^2\frac{\mathrm{d}^2Z}{\mathrm{d}x^2} + x\frac{\mathrm{d}Z}{\mathrm{d}x} + \left[x^2 - \left(n + \frac{1}{2}\right)^2\right]Z = 0 \qquad \text{[A1.33]}$$

1 The constant C is chosen since equation [A1.29] has no bounded solution and is therefore only physically admissible in the case where C is equal to $n(n+1)$.

Its solutions are Bessel functions of the first and second kind of fractional order:

$$Z_n(r) = a_n J_{n+\frac{1}{2}}(kr) + b_n Y_{n+\frac{1}{2}}(kr) \qquad [A1.34]$$

The $R_n(kr)$ functions are thus expressed by:

$$R_n(kr) = a'_n j_n(kr) + b'_n y_n(kr) \qquad [A1.35]$$

where the *spherical Bessel functions* of the first kind, j_n, and the second kind, y_n, are defined by:

$$j_n(x) = \sqrt{\frac{\pi}{2x}} J_{n+\frac{1}{2}}(x) \quad \text{and} \quad y_n(x) = \sqrt{\frac{\pi}{2x}} Y_{n+\frac{1}{2}}(x) \qquad [A1.36]$$

In some cases, it is preferable to introduce the spherical Hankel functions:

$$h_n^{(1)}(x) = j_n(x) + i y_n(x) \quad \text{and} \quad h_n^{(2)}(x) = j_n(x) - i y_n(x) \qquad [A1.37]$$

to express the function $R_n(kr)$:

$$R_n(kr) = a''_n h_n^{(1)}(kr) + b''_n h_n^{(2)}(kr) \qquad [A1.38]$$

A1.2.1. *Properties of spherical Bessel and Hankel functions*

The derivatives of the spherical Bessel and Hankel functions satisfy recurrence relations identical to [A1.14]:

$$\frac{\mathrm{d}f_n(x)}{\mathrm{d}x} = \frac{n}{x} f_n(x) - f_{n+1}(x) = \frac{1}{2}\left[f_{n-1}(x) - f_{n+1}(x)\right] \qquad [A1.39]$$

At $x = 0$, the function j_0 takes the value 1, while the other functions, j_n, are zero and the y_n functions are singular. It follows that the Hankel functions are also singular at the origin and their asymptotic behavior in the far field:

$$\lim_{x\to\infty} h_n^{(1)}(x) = (-i)^n \frac{e^{ix}}{ix} \quad \text{and} \quad \lim_{x\to\infty} h_n^{(2)}(x) = -i^n \frac{e^{-ix}}{ix} \qquad [A1.40]$$

is that of a divergent wave for $h_n^{(1)}$ and a convergent wave for $h_n^{(2)}$.

A1.2.2. *Properties of Legendre functions*

The associated Legendre functions verify the orthogonality relation:

$$\int_0^\pi P_n^m(\cos\theta)P_\nu^m(\cos\theta)\sin\theta\,\mathrm{d}\theta = \frac{2(n+m)!}{(2n+1)(n-m)!}\delta_{n\nu} \qquad \text{[A1.41]}$$

If m and μ are non-zero, they also verify the relation:

$$\int_0^\pi \frac{P_n^m(\cos\theta)}{\sin\theta}\frac{P_n^\mu(\cos\theta)}{\sin\theta}\sin\theta\,\mathrm{d}\theta = \frac{(n+m)!}{m(n-m)!}\delta_{m\mu} \qquad \text{[A1.42]}$$

Two other relations involving the Legendre functions are also useful:

$$\int_0^\pi \left[\frac{\mathrm{d}P_n^m(\cos\theta)}{\mathrm{d}\theta}\frac{\mathrm{d}P_\nu^m(\cos\theta)}{\mathrm{d}\theta} + m\frac{P_n^m(\cos\theta)}{\sin\theta}\frac{P_\nu^m(\cos\theta)}{\sin\theta}\right]\sin\theta\,\mathrm{d}\theta = \frac{2n(n+1)(n+m)!}{(2n+1)(n-m)!}\delta_{n,\nu} \qquad \text{[A1.43]}$$

and:

$$\int_0^\pi \left[\frac{\mathrm{d}P_n^m(\cos\theta)}{\mathrm{d}\theta}\frac{P_\nu^m(\cos\theta)}{\sin\theta} + \frac{\mathrm{d}P_\nu^m(\cos\theta)}{\mathrm{d}\theta}\frac{P_n^m(\cos\theta)}{\sin\theta}\right]\sin\theta\,\mathrm{d}\theta = 0 \;\; \forall m \neq 0 \qquad \text{[A1.44]}$$

A1.2.3. *Expansion on spherical harmonics*

Any scalar function that is a solution to the Helmholtz equation can be developed as an infinite sum of modes:

$$\phi(r,\theta,\varphi) = \sum_n^\infty \sum_{m=-n}^n \left[A_{n,m}h_n^{(1)}(kr) + B_{n,m}h_n^{(2)}(kr)\right] Y_{n,m}(\theta,\varphi) \qquad \text{[A1.45]}$$

called *normalized spherical harmonics*:

$$Y_{n,m}(\theta,\varphi) = \sqrt{\frac{2n+1}{4\pi}\frac{(n-m)!}{(n+m)!}}P_n^m(\cos\theta)e^{im\varphi} \qquad \text{[A1.46]}$$

The amplitudes $A_{n,m}$ of the spherical harmonics are deduced from the boundary conditions. The functions $Y_{n,m}(\theta, \varphi)$ are orthogonal and normalized functions; they verify the relation:

$$\int_0^{\pi} \int_0^{2\pi} Y_{n,m}(\theta, \varphi) Y^{\star}_{\nu,\mu}(\theta, \varphi) \sin\theta \, \mathrm{d}\theta \, \mathrm{d}\varphi = \delta_{n\nu}\delta_{m\mu} \qquad \text{[A1.47]}$$

A development in spherical harmonics, very useful for acoustic scattering problems, is that of a plane wave propagating in the z direction:

$$e^{ikr\cos\theta} = \sum_{n=0}^{\infty} i^n (2n+1) j_n(kr) P_n(\cos\theta) \qquad \text{[A1.48]}$$

Appendix 2

Matrix Representation of a Dioptric System

The objective of this time-harmonic representation is to express the amplitudes of bulk waves that propagate in a solid occupying the half-space $x_2 > 0$. These amplitudes are expressed as a function of the mechanical quantities that are conserved at an interface, that is, the three components of mechanical displacement $\underline{u}$ and those of mechanical traction: $T_i = \sigma_{ij}l_j = \sigma_{i2}$, for a surface of normal $\underline{l}$ parallel to the x_2 axis. The phenomena associated with the emission, detection and diffraction of elastic waves can be formulated using a mixed matrix $\boldsymbol{M}$, whereby the vector $\underline{b}$ for the amplitudes of the reflected (emitted) bulk waves and the mechanical amplitude $\underline{u}$ on the surface is related to the vector $\underline{a}$ for the amplitudes of the incident bulk waves and to the mechanical traction $\underline{T}$. This matrix $\boldsymbol{M}$ consists of four sub-matrices $\boldsymbol{D}$, $\boldsymbol{E}$, $\boldsymbol{R}$ and $\boldsymbol{Y}$:

$$\begin{pmatrix} \underline{b} \\ \underline{u} \end{pmatrix} = \begin{pmatrix} \boldsymbol{D} & \boldsymbol{E} \\ \boldsymbol{R} & \boldsymbol{Y} \end{pmatrix} \begin{pmatrix} \underline{a} \\ \underline{T} \end{pmatrix} \qquad \text{with} \qquad \boldsymbol{M} = \begin{pmatrix} \boldsymbol{D} & \boldsymbol{E} \\ \boldsymbol{R} & \boldsymbol{Y} \end{pmatrix} \qquad [A2.1]$$

Generally, the vectors $\underline{a}$ and $\underline{b}$ have three components (one for each bulk wave), and the vectors $\underline{u}$ and $\underline{T}$ also have three components:

$$\underline{u} = \begin{pmatrix} u_1 \\ u_2 \\ u_3 \end{pmatrix} \qquad \text{and} \qquad \underline{T} = \begin{pmatrix} \sigma_{12} \\ \sigma_{22} \\ \sigma_{32} \end{pmatrix} \qquad [A2.2]$$

The sub-matrices are thus square (3×3) matrices:

– the diffraction matrix $\boldsymbol{D}$ provides the amplitudes of reflected waves as functions of the amplitudes of the incident waves when the surface is free ($\underline{T} = \underline{0}$);

– the emission matrix $\boldsymbol{E}$ gives, in the absence of incident waves ($\underline{a} = \underline{0}$), the amplitudes $\underline{b}$ of bulk waves created by a uniform distribution of forces $\underline{T}$ on the surface;

– the reception matrix $\boldsymbol{R}$ is useful for calculating the mechanical displacement $\underline{u}$ induced, on the free surface ($\underline{T} = \underline{0}$), by bulk waves of amplitude $\underline{a}$;

– the admittance matrix $\boldsymbol{Y}$ gives, in the harmonic case, the mechanical displacement created by a uniform distribution of forces exerted on the surface. The components of its Fourier transform are the Green's functions of the half-space for mechanical surface phenomena.

A2.1. Diffraction matrix

Consider an incident, homogeneous plane wave of amplitude a, whose polarization is defined by the unit vector $\underline{p}$. For transversely independent problems, the wave vector $\underline{k}$ is contained in the x_1x_2 plane:

$$u_l = ap_l e^{i(k_1x_1+k_2x_2-\omega t)} \qquad \text{[A2.3]}$$

Upon omitting the propagation factor, the stresses σ_{i2} are in the form:

$$\sigma_{i2} = c_{i2kl}\frac{\partial u_l}{\partial x_k} = i\left(k_1c_{i21l} + k_2c_{i22l}\right)ap_l \qquad \text{[A2.4]}$$

By introducing the dimensionless parameters $m = k_2/k_1$, which defines the propagation direction in the x_1x_2 plane, the mechanical traction involved in the boundary conditions on the surface $x_2 = 0$ is given by:

$$T_i = \sigma_{i2} = ik_1A_ia \qquad \text{with} \qquad A_i = (c_{i21l} + mc_{i22l})p_l \qquad \text{[A2.5]}$$

The *mechanical traction* on the surface is the sum of those created by each incident wave (denoted by the index $I = 1, 2, 3$); by writing $t_i = T_i/ik_1$:

$$t_i = \sum_{I=1}^{3} A_{iI}a_I \qquad \longrightarrow \qquad \underline{t} = \boldsymbol{A}\underline{a} \qquad \text{[A2.6]}$$

where $\boldsymbol{A} = [A_{iI}]$ is the (3×3) matrix obtained by juxtaposing the column vectors A_I of each incident bulk wave. Identical considerations for the reflected waves (denoted by the index $J = 1, 2, 3$), of amplitude $\underline{b}$ and unit polarization $\underline{q}$, lead to the definition of the matrix $\boldsymbol{B}$ such that:

$$\boldsymbol{B} = [B_{iJ}] \qquad \text{with} \qquad B_{iJ} = (c_{i21l} + m_Jc_{i22l})q_{lJ} \qquad \text{[A2.7]}$$

The total mechanical traction is given by the sum:

$$\underline{t} = \boldsymbol{A}\underline{a} + \boldsymbol{B}\underline{b} \qquad \longrightarrow \qquad \underline{b} = -\boldsymbol{B}^{-1}\boldsymbol{A}\underline{a} + \boldsymbol{B}^{-1}\underline{t} \tag{A2.8}$$

Since $\underline{t} = \underline{0}$ on the free surface $x_2 = 0$, the amplitudes of the reflected bulk waves are expressed as functions of the amplitudes of the incident waves by:

$$\underline{b} = \boldsymbol{D}\underline{a}, \qquad \text{with} \qquad \boldsymbol{D} = -\boldsymbol{B}^{-1}\boldsymbol{A} \tag{A2.9}$$

The nine components of the diffraction matrix $\boldsymbol{D}$ are the *reflection coefficients*. For example, by attributing the index 1 to the quasi-longitudinal (QL) wave and the indices 2 and 3 to the quasi-transverse (QT) waves, the vector $\underline{a} = (a_1, 0, 0)^t$ represents an incident QL wave of amplitude a_1. The components of the first row in the matrix $\boldsymbol{D}$ are the reflection coefficients into the QL wave ($r_{11} = b_1/a_1$) and the conversion coefficient into the QT wave ($r_{12} = b_2/a_1$, $r_{13} = b_3/a_1$). According to the matrix inversion rule:

$$\boldsymbol{D} = -\boldsymbol{B}^{-1}\boldsymbol{A} = -\frac{\mathrm{Adj}[\boldsymbol{B}]}{\det[\boldsymbol{B}]}\boldsymbol{A} \tag{A2.10}$$

the denominator for all the reflection coefficients is the determinant of the matrix $\boldsymbol{B}$.

In the absence of any incident wave ($\underline{a} = \underline{0}$), the homogeneous system $\boldsymbol{B}\underline{b} = \underline{0}$ has a non-trivial solution if:

$$\det[\boldsymbol{B}] = 0 \qquad \longrightarrow \qquad \underline{b} \neq \underline{0} \tag{A2.11}$$

An elastic wave with non-zero displacement $\underline{b}$ can propagate on the free surface of a solid. This wave was discovered in the case of an isotropic solid by Lord Rayleigh in 1885. The characteristic equation of a Rayleigh wave is therefore obtained by canceling the denominator of the reflection coefficients on the free surface of an isotropic solid. This observation can also be applied to waves that propagate along the interface between a fluid and a solid, or between two solids, whether isotropic or anisotropic.

A2.2. Emission, admittance and reception matrices

In the absence of any incident wave ($\underline{a} = \underline{0}$) and according to relation [A2.8]: $\underline{b} = \boldsymbol{B}^{-1}\underline{t}$, the amplitudes $\underline{b}$ of the emitted *bulk waves* are expressed as functions of the stress vector $\underline{T}$ on the surface, using the emission matrix $\boldsymbol{E}$:

$$\underline{b} = \boldsymbol{E}\underline{T} \qquad \text{with} \qquad ik_1\boldsymbol{E} = \boldsymbol{B}^{-1} \tag{A2.12}$$

With $\boldsymbol{Q} = [q_{lJ}]$ denoting the polarization matrix of the emitted (reflected) waves, the mechanical displacement on the surface:

$$u_l = \sum_{I=1}^{3} q_{lI} b_I \qquad \text{hence} \qquad \underline{u} = \boldsymbol{Q}\underline{b} \qquad \text{[A2.13]}$$

can be expressed as a function of the stress vector $\underline{T}$ using the admittance matrix $\boldsymbol{Y}$:

$$\underline{u} = \boldsymbol{Y}\underline{T} \qquad \text{with} \qquad \boldsymbol{Y} = \boldsymbol{Q}\boldsymbol{E} \qquad \text{[A2.14]}$$

By using $\boldsymbol{P} = [p_{lI}]$ to denote the polarization matrix of incident waves, the total mechanical displacement on the surface created by the incident waves of amplitude $\underline{a}$, and reflected waves of amplitude $\underline{b} = -\boldsymbol{B}^{-1}\boldsymbol{A}\underline{a}$, is expressed using the reception matrix $\boldsymbol{R}$:

$$\underline{u} = \boldsymbol{P}\underline{a} + \boldsymbol{Q}\underline{b} = \boldsymbol{R}\underline{a} \qquad \text{with} \qquad \boldsymbol{R} = \boldsymbol{P} - \boldsymbol{Q}\boldsymbol{B}^{-1}\boldsymbol{A} \qquad \text{[A2.15]}$$

A2.3. Isotropic solid

Given that only two bulk waves can propagate in an isotropic solid and for the displacements $\underline{u}$ and mechanical tractions $\underline{T}$ contained in the (x_1, x_2) plane, the sub-matrices of the mixed matrix are (2×2) matrices. According to the Snell–Descartes law, the component s_1 of the slowness vector is identical for all waves and $s_3 = 0$. The normal components s_{2L} and s_{2T} are given by the dispersion relations:

$$s_1^2 + s_{2L}^2 = s_L^2 \qquad \text{and} \qquad s_1^2 + s_{2T}^2 = s_T^2 \qquad \text{[A2.16]}$$

where $s_L = 1/V_L$ and $s_T = 1/V_T$ are the slownesses of longitudinal and transverse waves. The development of relation [A2.4] for the emitted waves, of amplitude $\underline{b}$ and of polarization vector $\underline{q}$:

$$T_i = \sigma_{i2} = i\omega(s_1 c_{i21l} + s_2 c_{i22l}) q_l b \qquad \text{[A2.17]}$$

with $C_{22} = C_{11}$ and $C_{21} = C_{11} - 2C_{66}$, leads to:

$$\begin{cases} T_1 = i\omega C_{66}(s_1 q_2 + s_2 q_1) b \\ T_2 = i\omega \left[C_{11}(s_1 q_1 + s_2 q_2) - 2C_{66} s_1 q_1\right] b \end{cases} \qquad \text{[A2.18]}$$

For the *longitudinal wave*, the normalized components $(q_1^2 + q_2^2 = 1)$ of the polarization vector are:

$$q_{1L} = s_1 V_L \qquad \text{and} \qquad q_{2L} = s_{2L} V_L \qquad \text{[A2.19]}$$

Using the relation $C_{11}(s_1^2 + s_{2L}^2) = \rho = C_{66}s_T^2$, the components of mechanical traction $\underline{T}_L$ are given by:

$$\begin{cases} T_{1L} = 2i\omega C_{66}s_1 s_{2L} V_L b_L \\ T_{2L} = i\omega \left[C_{11}(s_1^2 + s_{2L}^2) - 2C_{66}s_1^2\right] V_L b_L = i\omega C_{66}(s_T^2 - 2s_1^2) V_L b_L \end{cases} \quad \text{[A2.20]}$$

Similarly, for the *transverse wave*:

$$q_{1T} = -s_{2T}V_T \qquad \text{and} \qquad q_{2T} = s_1 V_T \quad \text{[A2.21]}$$

the components of the mechanical traction $\underline{T}_T$ are written as:

$$\begin{cases} T_{1T} = i\omega C_{66}(s_1^2 - s_{2T}^2)V_T b_T = i\omega C_{66}(2s_1^2 - s_T^2)V_T b_T \\ T_{2T} = 2i\omega C_{66}s_1 s_{2T} V_T b_T \end{cases} \quad \text{[A2.22]}$$

By introducing the Lamé constant $\mu = C_{66}$, the stress vector $\underline{T} = \underline{T}_L + \underline{T}_T$ is written in matrix form as:

$$\begin{bmatrix} T_1 \\ T_2 \end{bmatrix} = i\omega\mu \begin{bmatrix} 2s_1 s_{2L} V_L & (2s_1^2 - s_T^2)V_T \\ (s_T^2 - 2s_1^2)V_L & 2s_1 s_{2T} V_T \end{bmatrix} \begin{bmatrix} b_L \\ b_T \end{bmatrix} \quad \text{[A2.23]}$$

The inversion leads to the emission matrix:

$$\boldsymbol{E} = \frac{1}{i\omega\mu\Delta} \begin{bmatrix} 2s_1 s_{2T} s_L & (s_T^2 - 2s_1^2)s_L \\ (2s_1^2 - s_T^2)s_T & 2s_1 s_{2L} s_T \end{bmatrix} \quad \text{[A2.24]}$$

where Δ is the Rayleigh determinant:

$$\Delta = (2s_1^2 - s_T^2)^2 + 4s_1^2 s_{2L} s_{2T} \quad \text{[A2.25]}$$

Considering relations [A2.19] and [A2.21], the polarization matrix of the reflected or emitted waves is:

$$\boldsymbol{Q} = \begin{bmatrix} q_{1L} & q_{1T} \\ q_{2L} & q_{2T} \end{bmatrix} = \begin{bmatrix} s_1 V_L & -s_{2T} V_T \\ s_{2L} V_L & s_1 V_T \end{bmatrix} \quad \text{[A2.26]}$$

and the admittance matrix $\boldsymbol{Y}$ can be deduced from relation [A2.14] $\boldsymbol{Y} = \boldsymbol{QE}$, with:

$$\begin{aligned} \boldsymbol{Y} &= \frac{1}{i\omega\mu\Delta}\begin{bmatrix} s_T^2 s_{2T} & -s_1\left(2s_1^2 - s_T^2 + 2s_{2L}s_{2T}\right) \\ s_1\left(2s_1^2 - s_T^2 + 2s_{2L}s_{2T}\right) & s_T^2 s_{2L} \end{bmatrix} \\ &= \frac{\boldsymbol{N}}{i\omega\mu\Delta} \end{aligned} \quad [A2.27]$$

References

Chapter 1

Achenbach, J. (1987). *Wave Propagation in Elastic Solids*. North-Holland, Amsterdam.

Bochud, N., Laurent, J., Bruno, F., Royer, D., Prada, C. (2018). Towards real-time assessment of anisotropic plate properties using elastic guided waves. *J. Acoust. Soc. Am.*, 143, 1138–1147.

Bruneau, M. (2006). *Fundamentals of Acoustics*. ISTE Ltd, London.

Corbel, C., Guillois, F., Royer, D., Fink, M.A., De Mol, R. (1993). Laser-generated elastic waves in carbon-epoxy composite. *IEEE Trans. Ultrason. Ferr.*, 40(6), 710–716.

Fink, M.A. and Cardoso, J.F. (1984). Diffraction effects in pulse-echo measurement. *IEEE Trans. Son. Ultrason.*, 31(4), 313–329.

Kino, G.S. (1987). *Acoustic Waves: Devices, Imaging, and Analog Signal Processing*. Prentice-Hall, Upper Saddle River.

Lamb, H. (1904). On the propagation of tremors over the surface of an elastic solid. *Phil. Trans. Roy. Soc. (London)*, A203, 1–42.

Maris, H.J. (1971). Enhancement of heat pulses in crystals due to elastic anisotropy. *J. Acoust. Soc. Am.*, 50(3B), 812–818.

Penttinen, A. and Luukkala, M. (1976). The impulse response and pressure nearfield of a curved ultrasonic radiator. *J. Phys. D: Appl. Phys.*, 9(10), 1547–1557.

Poncelet, O. and Deschamps, M. (2009). The Cagniard–de Hoop method. In *Materials and Acoustics Handbook*, Bruneau, M. and Potel, C. (eds). ISTE Ltd, London and John Wiley & Sons, New York.

Royer, D. (2001). Mixed matrix formulation for the analysis of laser-generated acoustic waves by a thermoelastic line source. *Ultrasonics*, 39(5), 345–354.

Stepanishen, P.R. (1971). Transient radiation from pistons in an infinite planar baffle. *J. Acoust. Soc. Am.*, 49(5B), 1629–1638.

Weaver, R.L., Sachse, W., Kim, K.Y. (1996). Transient elastic waves in a transversely isotropic plate. *J. Appl. Mech.*, 63(2), 337–346.

Chapter 2

Brillouin, L. (1949). The scattering cross section of spheres for electromagnetic waves. *J. Appl. Phys.*, 20(11), 1110–1125.

Conoir, J.-M. and Norris, A.N. (2010). Effective wavenumbers and reflection coefficients for an elastic medium containing random configurations of cylindrical scatterers. *Wave Motion*, 47(3), 183–197.

Derode, A., Mamou, V., Tourin, A. (2006). Influence of correlations between scatterers on the attenuation of the coherent wave in a random medium. *Phys. Rev. E*, 74, 036606.

Duranteau, M., Valier-Brasier, T., Conoir, J.-M., Wunenburger, R. (2016). Random acoustic metamaterial with a subwavelength dipolar resonance. *J. Acoust. Soc. Am.*, 139(6), 3341–3352.

Einspruch, N.G., Witterholt, E.J., Truell, R. (1960). Scattering of a plane transverse wave by a spherical obstacle in an elastic medium. *J. Appl. Phys.*, 31(5), 806–818.

Faran, J.J. (1951). Sound scattering by solid cylinders and spheres. *J. Acoust. Soc. Am.*, 23, 405–418.

Fikioris, J.G. and Waterman, P.C. (1964). Multiple scattering of waves II. "Hole corrections" in the scalar case. *J. Math. Phys.*, 5, 1413–1420.

Foldy, L.L. (1945). The multiple scattering of waves I. General theory of isotropic scattering by randomly distributed scatterers. *Phys. Rev.*, 67(3–4), 107–119.

Gaunaurd, G.C. and Überall, H. (1978). Theory of resonant scattering from spherical cavities in elastic and viscoelastic media. *J. Acoust. Soc. Am.*, 63(6), 1699–1712.

Ishimaru, A. (1978). *Wave Propagation and Scattering in Random Media, volume 2: Multiple Scattering, Turbulence, Rough Surfaces, and Remote Sensing*. Academic Press, Cambridge.

Lax, M. (1952). Multiple scattering of waves II. The effective field in dense systems. *Phys. Rev.*, 85, 621–629.

Linton, C.M. and Martin, P.A. (2005). Multiple scattering by random configurations of circular cylinders: Second-order corrections for the effective wavenumber. *J. Acoust. Soc. Am.*, 117(6), 3413–3423.

Liu, Y., Wu, R.-S., Ying, C.F. (2000). Scattering of elastic waves by an elastic or viscoelastic cylinder. *Geophys. J. Int.*, 142, 439–460.

Luppé, F., Conoir, J.-M., Norris, A.N. (2012). Effective wave numbers for thermo-viscoelastic media containing random configurations of spherical scatterers. *J. Acoust. Soc. Am.*, 131(2), 1113–1120.

Morse, P.M. and Feshbach, H. (1953). *Methods of Theoretical Physics, Part II*. McGraw-Hill, New York.

Rohfritsch, A., Conoir, J.-M., Marchiano, R., Valier-Brasier, T. (2019). Numerical simulation of two-dimensional multiple scattering of sound by a large number of circular cylinders. *J. Acoust. Soc. Am.*, 145(6), 3320–3329.

Tsang, L. and Kong, J. (1982). Effective propagation constants for coherent electromagnetic wave propagation in media embedded with dielectric scatters. *J. Appl. Phys.*, 53, 7162.

Valier-Brasier, T., Conoir, J.-M., Coulouvrat, F., Thomas, J.-L. (2015). Sound propagation in dilute suspensions of spheres: Analytical comparison between coupled phase model and multiple scattering theory. *J. Acoust. Soc. Am.*, 138(4), 2598–2612.

Waterman, P.C. and Truell, R. (1961). Multiple scattering of waves. *J. Math. Phys.*, 2(4), 512–537.

Chapter 3

Achenbach, J. (1987). *Wave Propagation in Elastic Solids*. North-Holland, Amsterdam.

Aki, K. and Richards, P.G. (1997). *Quantitative Seismology*, 2nd edition. University Science Books, Melville.

Carslaw, H.S. and Jaeger, J.C. (1959). *Conduction of Heat in Solids*. Clarendon Press, Oxford.

Chen, D.-P. and Haus, H.A. (1985). Analysis of metal-strip saw gratings and transducers. *IEEE Trans. Son. Ultrason.*, 32(3), 395–408.

Datta, S. and Hunsinger, B.J. (1979). First-order reflection coefficient of surface acoustic waves from thin-strip overlays. *J. Appl. Phys.*, 50(9), 5661–5665.

Datta, S. and Hunsinger, B.J. (1980). An analytical theory for the scattering of surface acoustic waves by a single electrode in a periodic array on a piezoelectric substrate. *J. Appl. Phys.*, 51(9), 4817–4823.

Dewhurst, R.J., Hutchins, D.A., Palmer, S.B., Scruby, C.B. (1982). Quantitative measurements of laser-generated acoustic waveforms. *J. Appl. Phys.*, 53(6), 4064–4071.

Hartmann, C.S., Bell, D.T., Rosenfeld, R.C. (1973). Impulse model design of acoustic surface-wave filters. *IEEE Trans. Microw. Theory,* 21(4), 162–175.

Hartmann, C.S., Wright, P.V., Kansy, R.J., Garber, E.M. (1982). An analysis of saw interdigital transducers with internal reflections and the application to the design of single-phase unidirectional transducers. *IEEE Ultrasonics Symp. Proc.*, 1, 40–45.

Higuet, J., Valier-Brasier, T., Dehoux, T., Audoin, B. (2011). Beam distortion detection and deflectometry measurements of gigahertz surface acoustic waves. *Rev. Sci. Instrum.*, 82(11), 114905.

Hutchins, D.A. (1988). Ultrasonic generation by pulsed laser. In *Physical Acoustics*, volume 18, Thurston R.N. and Pierce, A.D. (eds). Academic Press Inc., Cambridge.

Krimholtz, R., Leedom, D.A., Matthaei, G.L. (1970). New equivalent circuits for elementary piezoelectric transducers. *Electron. Lett.*, 6(13), 398–399.

Kwaaitaal, T. (1974). Contribution to the interferometric measurement of sub-angstrom vibrations. *Rev. Sci. Instrum.*, 45(1), 39–41.

Lewis, M.F. (1983). A different approach to the wave-scattering properties of interdigital transducers. *IEEE Trans. Son. Ultrason.*, 30(1), 55–56.

Mason, W.P. (1948). *Electromechanical Transducers and Wave Filters*, 2nd edition. Van Nostrand-Reinhold, Princeton.

Monchalin, J. (1986). Optical detection of ultrasound. *IEEE Trans. Ultrason. Ferr.*, 33(5), 485–499.

Ohigashi, H., Koga, K., Suzuki, M., Nakanishi, T., Kimura, K., Hashimoto, N. (1984). Piezoelectric and ferroelectric properties of P(VDF-TrFE) copolymers and their application to ultrasonic transducers. *Ferroelectrics,* 60(1), 263–276.

Ready, J.F. (1971). *Effect of High Power Radiation.* Academic Press, New York.

Redwood, M. (1961). Transient performance of a piezoelectric transducer. *J. Acoust. Soc. Am.*, 33(4), 527–536.

Rose, L.R.F. (1984). Point-source representation for laser-generated ultrasound. *J. Acoust. Soc. Am.*, 75(3), 723–732.

Royer, D. (2001). Mixed matrix formulation for the analysis of laser-generated acoustic waves by a thermoelastic line source. *Ultrasonics,* 39(5), 345–354.

Royer, D. and Dieulesaint, E. (1986). Optical detection of sub-angstrom transient mechanical displacements. *IEEE Ultrason. Symp. Proc.*, 1, 527–530.

Royer, D. and Dieulesaint, E. (1999). *Ondes élastiques dans les solides : génération, interaction acousto-optique, applications*, Volume 2. Masson, Paris.

Scruby, C.B. and Drain, L.E. (1990). *Laser Ultrasonics Techniques and Applications.* Adam Hilger, London.

Smith, W.A. and Auld, B.A. (1991). Modeling 1-3 composite piezoelectrics: Thickness-mode oscillations. *IEEE Trans. Ultrason. Ferr.*, 38(1), 40–47.

Tancrell, R.H. and Holland, M.G. (1971). Acoustic surface wave filters. *Proceedings of the IEEE,* 59(3), 393–409.

Thompson, R.B. (1990). Physical principles of measurements with EMAT transducers. In *Physical Acoustics,* Volume 19, Thurston, R.N. and Pierce, A.D. (eds). Academic Press Inc., Cambridge.

Thomsen, C., Strait, J., Vardeny, Z., Maris, H.J., Tauc, J., Hauser, J.J. (1984). Coherent phonon generation and detection by picosecond light pulses. *Phys. Rev. Lett.,* 53, 989–992.

White, R.M. and Voltmer, F.W. (1965). Direct piezoelectric coupling to surface elastic waves. *Appl. Phys. Lett.,* 7(12), 314–316.

Wright, P.V. (1989). A new generalized modeling of saw transducers and gratings. In *Proceedings of the 43rd Annual Symposium on Frequency Control,* IEEE, Denver, 596–605.

General bibliography

Achenbach, J. (2012). *Wave Propagation in Elastic Solids.* North-Holland, New York.

Aki, K. and Richards, P.G. (2002). *Quantitative Seismology.* University Science Books, Herndon.

Auld, B.A. (1990). *Acoustic Fields and Waves in Solids,* Volume 1. R.E. Krieger Publishing Company, Malabar.

Brekhovskikh, L.M. and Godin, O.A. (1998). *Acoustics of Layered Media I: Plane and Quasi-Plane Waves,* 2nd edition. Springer, Berlin.

Bruneau, M. and Potel, C. (2009). *Materials and Acoustics Handbook.* ISTE Ltd, London and John Wiley & Sons, New York.

Campbell, C. (1989). *Surface Acoustic Waves Devices and Their Signal Processing Applications.* Academic Press, Boston.

Cheeke, J.D.N. (2002). *Fundamentals and Applications of Ultrasonic Waves.* CRC Press, Boca Raton.

Elmore, W.C. and Heald, M.A. (1985). *Physics of Waves.* Dover Publications, New York.

Gusev, V. and Karabutov, A. (1993). *Laser Optoacoustics.* AIP Press, College Park.

Hutchins, D.A. and Hayward, G. (1990). Radiated fields of ultrasonic transducers. *Physical Acoustics,* 19, 1–80.

Kennett, B. (1983). *Seismic Wave Propagation in Stratified Media.* Cambridge University Press, Cambridge.

Kino, G.S. (1987). *Acoustic Waves: Devices, Imaging, and Analog Signal Processing.* Prentice-Hall, Upper Saddle River.

Martin, P.A. (2006). *Multiple Scattering – Interaction of Time-Harmonic Waves with N Obstacles.* Cambridge University Press, Cambridge.

Miklowitz, J. (2012). *The Theory of Elastic Waves and Waveguides*. North-Holland, New York.

Mindlin, R.D. and Yang, J. (2006). *An Introduction to the Mathematical Theory of Vibrations of Elastic Plates*. World Scientific, Singapore.

Morgan, D.P. (1991). *Surface-Wave Devices for Signal Processing*. Elsevier, New York.

Nayfeh, A.H. (1995). *Wave Propagation in Layered Anisotropic Media*. North-Holland, New York.

Ristic, V.M. (1983). *Principles of Acoustic Devices*. John Wiley & Sons, New York.

Rose, J.L. (2004). *Ultrasonic Waves in Solid Media*. Cambridge University Press, Cambridge.

Rose, J.L. (2014). *Ultrasonic Guided Waves in Solid Media*. Cambridge University Press, Cambridge.

Royer, D. and Valier-Brasier, T. (2022). *Elastic Waves in Solids 1: Propagation*. ISTE Ltd, London and John Wiley & Sons, New York.

Royer, D. and Dieulesaint, E. (1996). *Ondes élastiques dans les solides : propagation libre et guidée*, Volume 1. Masson, Paris.

Royer, D. and Dieulesaint, E. (1999). *Ondes élastiques dans les solides : génération, interaction acousto-optique, applications*, Volume 2. Masson, Paris.

Salençon, J. (1988). *Mécanique des milieux continus II – Élasticité-milieux curvilignes*. Ellipses, Paris.

Index

Other titles from

iSTE

in

Waves

2022

ROYER Daniel, VALIER-BRASIER Tony
Elastic Waves in Solids 1: Propagation

2021

DAHOO Pierre-Richard, LAKHLIFI Azzedine
Infrared Spectroscopy of Symmetric and Spherical Top Molecules for Space Observation 1 (Infared Spectroscopy Set – Volume 3)

Infrared Spectroscopy of Symmetric and Spherical Top Molecules for Space Observation 2 (Infrared Spectroscopy Set – Volume 4)

FAVENNEC Pierre-Noël
Electromagnetic Waves 1: Maxwell's Equations, Wave Propagation
Electromagnetic Waves 2: Antennas

LACAZE Pierre Camille, LACROIX Jean-Christophe
Nanotechnology and Nanomaterials for Energy

RÉVEILLAC Jean-Michel
Recording and Voice Processing 1: History and Generalities
Recording and Voice Processing 2: Working in the Studio

SAKHO Ibrahima
Nuclear Physics 1: Nuclear Deexcitations, Spontaneous Nuclear Reactions

2020

DANIELE Vito G., LOMBARDI Guido
Scattering and Diffraction by Wedges 1: The Wiener-Hopf Solution - Advanced Applications (Waves and Scattering Set – Volume 1)

Scattering and Diffraction by Wedges 2: The Wiener-Hopf Solution - Advanced Applications (Waves and Scattering Set – Volume 2)

SAKHO Ibrahima
Introduction to Quantum Mechanics 2: Wave-Corpuscle, Quantization & Schrödinger's Equation

2019

BERTRAND Pierre, DEL SARTO Daniele, GHIZZO Alain
The Vlasov Equation 1: History and General Properties

DAHOO Pierre-Richard, LAKHLIFI Azzedine
Infrared Spectroscopy of Triatomics for Space Observation
(Infrared Spectroscopy Set – Volume 2)

RÉVEILLAC Jean-Michel
Electronic Music Machines: The New Musical Instruments

ROMERO-GARCIA Vicente, HLADKY-HENNION Anne-Christine
Fundamentals and Applications of Acoustic Metamaterials: From Seismic to Radio Frequency
(Metamaterials Applied to Waves Set – Volume 1)

SAKHO Ibrahima
Introduction to Quantum Mechanics 1: Thermal Radiation and Experimental Facts Regarding the Quantization of Matter

2018

SAKHO Ibrahima
Screening Constant by Unit Nuclear Charge Method: Description and Application to the Photoionization of Atomic Systems

2017

DAHOO Pierre-Richard, LAKHLIFI Azzedine
Infrared Spectroscopy of Diatomics for Space Observation (Infrared Spectroscopy Set – Volume 1)

PARET Dominique, HUON Jean-Paul
Secure Connected Objects

PARET Dominque, SIBONY Serge
Musical Techniques: Frequencies and Harmony

RÉVEILLAC Jean-Michel
Analog and Digital Sound Processing

STAEBLER Patrick
Human Exposure to Electromagnetic Fields

2016

ANSELMET Fabien, MATTEI Pierre-Olivier
Acoustics, Aeroacoustics and Vibrations

BAUDRAND Henri, TITAOUINE Mohammed, RAVEU Nathalie
The Wave Concept in Electromagnetism and Circuits: Theory and Applications

PARET Dominique
Antennas Designs for NFC Devices

PARET Dominique
Design Constraints for NFC Devices

WIART Joe
Radio-Frequency Human Exposure Assessment

2015

PICART Pascal
New Techniques in Digital Holography

2014

APPRIOU Alain
Uncertainty Theories and Multisensor Data Fusion

JARRY Pierre, BENEAT Jacques N.
RF and Microwave Electromagnetism

LAHEURTE Jean-Marc
UHF RFID Technologies for Identification and Traceability

SAVAUX Vincent, LOUËT Yves
MMSE-based Algorithm for Joint Signal Detection, Channel and Noise Variance Estimation for OFDM Systems

THOMAS Jean-Hugh, YAAKOUBI Nourdin
New Sensors and Processing Chain

TING Michael
Molecular Imaging in Nano MRI

VALIÈRE Jean-Christophe
Acoustic Particle Velocity Measurements using Laser: Principles, Signal Processing and Applications

VANBÉSIEN Olivier, CENTENO Emmanuel
Dispersion Engineering for Integrated Nanophotonics

2013

BENMAMMAR Badr, AMRAOUI Asma
Radio Resource Allocation and Dynamic Spectrum Access

BOURLIER Christophe, PINEL Nicolas, KUBICKÉ Gildas
Method of Moments for 2D Scattering Problems: Basic Concepts and Applications

GOURE Jean-Pierre
Optics in Instruments: Applications in Biology and Medicine

LAZAROV Andon, KOSTADINOV Todor Pavlov
Bistatic SAR/GISAR/FISAR Theory Algorithms and Program Implementation

LHEURETTE Eric
Metamaterials and Wave Control

PINEL Nicolas, BOURLIER Christophe
Electromagnetic Wave Scattering from Random Rough Surfaces: Asymptotic Models

SHINOHARA Naoki
Wireless Power Transfer via Radiowaves

TERRE Michel, PISCHELLA Mylène, VIVIER Emmanuelle
Wireless Telecommunication Systems

2012

LALAUZE René
Chemical Sensors and Biosensors

LE MENN Marc
Instrumentation and Metrology in Oceanography

LI Jun-chang, PICART Pascal
Digital Holography

2011

BECHERRAWY Tamer
Mechanical and Electromagnetic Vibrations and Waves

BESNIER Philippe, DÉMOULIN Bernard
Electromagnetic Reverberation Chambers

GOURE Jean-Pierre
Optics in Instruments

GROUS Ammar
Applied Metrology for Manufacturing Engineering

LE CHEVALIER François, LESSELIER Dominique, STARAJ Robert
Non-standard Antennas

2010

BEGAUD Xavier
Ultra Wide Band Antennas

MARAGE Jean-Paul, MORI Yvon
Sonar and Underwater Acoustics

2009

BOUDRIOUA Azzedine
Photonic Waveguides

BRUNEAU Michel, POTEL Catherine
Materials and Acoustics Handbook

DE FORNEL Frédérique, FAVENNEC Pierre-Noël
Measurements using Optic and RF Waves

FRENCH COLLEGE OF METROLOGY
Transverse Disciplines in Metrology

2008

FILIPPI Paul J.T.
Vibrations and Acoustic Radiation of Thin Structures

LALAUZE René
Physical Chemistry of Solid-Gas Interfaces

2007

KUNDU Tribikram
Advanced Ultrasonic Methods for Material and Structure Inspection

PLACKO Dominique
Fundamentals of Instrumentation and Measurement

RIPKA Pavel, TIPEK Alois
Modern Sensors Handbook

2006

BALAGEAS Daniel *et al.*
Structural Health Monitoring

BOUCHET Olivier *et al.*
Free-Space Optics

BRUNEAU Michel, SCELO Thomas
Fundamentals of Acoustics

FRENCH COLLEGE OF METROLOGY
Metrology in Industry

GUILLAUME Philippe
Music and Acoustics

GUYADER Jean-Louis
Vibration in Continuous Media

Printed and bound by CPI Group (UK) Ltd, Croydon, CR0 4YY

15/09/2026

14962347-0003